LONDON MATHEMATICAL SOCIETY LECTURE NOTE SERIES

Managing Editor: Professor J.W.S. Cassels, Department of Pure Mathematics and Mathematical Statistics, University of Cambridge, 16 Mill Lane, Cambridge CB2 1SB, England

The books in the series listed below are available from booksellers, or, in case of difficulty, from Cambridge University Press.

34 Representation theory of Lie groups, M.F. ATIYAH *et al*
36 Homological group theory, C.T.C. WALL (ed)
39 Affine sets and affine groups, D.G. NORTHCOTT
40 Introduction to H_p spaces, P.J. KOOSIS
43 Graphs, codes and designs, P.J. CAMERON & J.H. VAN LINT
45 Recursion theory: its generalisations and applications, F.R. DRAKE & S.S. WAINER (eds)
46 p-adic analysis: a short course on recent work, N. KOBLITZ
49 Finite geometries and designs, P. CAMERON, J.W.P. HIRSCHFELD & D.R. HUGHES (eds)
50 Commutator calculus and groups of homotopy classes, H.J. BAUES
51 Synthetic differential geometry, A. KOCK
57 Techniques of geometric topology, R.A. FENN
59 Applicable differential geometry, M. CRAMPIN & F.A.E. PIRANI
62 Economics for mathematicians, J.W.S. CASSELS
66 Several complex variables and complex manifolds II, M.J. FIELD
69 Representation theory, I.M. GELFAND *et al*
74 Symmetric designs: an algebraic approach, E.S. LANDER
76 Spectral theory of linear differential operators and comparison algebras, H.O. CORDES
77 Isolated singular points on complete intersections, E.J.N. LOOIJENGA
78 A primer on Riemann surfaces, A.F. BEARDON
79 Probability, statistics and analysis, J.F.C. KINGMAN & G.E.H. REUTER (eds)
80 Introduction to the representation theory of compact and locally compact groups, A. ROBERT
81 Skew fields, P.K. DRAXL
82 Surveys in combinatorics, E.K. LLOYD (ed)
83 Homogeneous structures on Riemannian manifolds, F. TRICERRI & L. VANHECKE
86 Topological topics, I.M. JAMES (ed)
87 Surveys in set theory, A.R.D. MATHIAS (ed)
88 FPF ring theory, C. FAITH & S. PAGE
89 An F-space sampler, N.J. KALTON, N.T. PECK & J.W. ROBERTS
90 Polytopes and symmetry, S.A. ROBERTSON
91 Classgroups of group rings, M.J. TAYLOR
92 Representation of rings over skew fields, A.H. SCHOFIELD
93 Aspects of topology, I.M. JAMES & E.H. KRONHEIMER (eds)
94 Representations of general linear groups, G.D. JAMES
95 Low-dimensional topology 1982, R.A. FENN (ed)
96 Diophantine equations over function fields, R.C. MASON
97 Varieties of constructive mathematics, D.S. BRIDGES & F. RICHMAN
98 Localization in Noetherian rings, A.V. JATEGAONKAR
99 Methods of differential geometry in algebraic topology, M. KAROUBI & C. LERUSTE
100 Stopping time techniques for analysts and probabilists, L. EGGHE
101 Groups and geometry, ROGER C. LYNDON
103 Surveys in combinatorics 1985, I. ANDERSON (ed)
104 Elliptic structures on 3-manifolds, C.B. THOMAS
105 A local spectral theory for closed operators, I. ERDELYI & WANG SHENGWANG
106 Syzygies, E.G. EVANS & P. GRIFFITH

107 Compactification of Siegel moduli schemes, C-L. CHAI
108 Some topics in graph theory, H.P. YAP
109 Diophantine analysis, J. LOXTON & A. VAN DER POORTEN (eds)
110 An introduction to surreal numbers, H. GONSHOR
111 Analytical and geometric aspects of hyperbolic space, D.B.A.EPSTEIN (ed)
112 Low-dimensional topology and Kleinian groups, D.B.A. EPSTEIN (ed)
113 Lectures on the asymptotic theory of ideals, D. REES
114 Lectures on Bochner-Riesz means, K.M. DAVIS & Y-C. CHANG
115 An introduction to independence for analysts, H.G. DALES & W.H. WOODIN
116 Representations of algebras, P.J. WEBB (ed)
117 Homotopy theory, E. REES & J.D.S. JONES (eds)
118 Skew linear groups, M. SHIRVANI & B. WEHRFRITZ
119 Triangulated categories in the representation theory of finite-dimensional algebras, D. HAPPEL
121 Proceedings of *Groups - St Andrews 1985*, E. ROBERTSON & C. CAMPBELL (eds)
122 Non-classical continuum mechanics, R.J. KNOPS & A.A. LACEY (eds)
124 Lie groupoids and Lie algebroids in differential geometry, K. MACKENZIE
125 Commutator theory for congruence modular varieties, R. FREESE & R. MCKENZIE
126 Van der Corput's method of exponential sums, S.W. GRAHAM & G. KOLESNIK
127 New directions in dynamical systems, T.J. BEDFORD & J.W. SWIFT (eds)
128 Descriptive set theory and the structure of sets of uniqueness, A.S. KECHRIS & A. LOUVEAU
129 The subgroup structure of the finite classical groups, P.B. KLEIDMAN & M.W.LIEBECK
130 Model theory and modules, M. PREST
131 Algebraic, extremal & metric combinatorics, M-M. DEZA, P. FRANKL & I.G. ROSENBERG (eds)
132 Whitehead groups of finite groups, ROBERT OLIVER
133 Linear algebraic monoids, MOHAN S. PUTCHA
134 Number theory and dynamical systems, M. DODSON & J. VICKERS (eds)
135 Operator algebras and applications, 1, D. EVANS & M. TAKESAKI (eds)
136 Operator algebras and applications, 2, D. EVANS & M. TAKESAKI (eds)
137 Analysis at Urbana, I, E. BERKSON, T. PECK, & J. UHL (eds)
138 Analysis at Urbana, II, E. BERKSON, T. PECK, & J. UHL (eds)
139 Advances in homotopy theory, S. SALAMON, B. STEER & W. SUTHERLAND (eds)
140 Geometric aspects of Banach spaces, E.M. PEINADOR and A. RODES (eds)
141 Surveys in combinatorics 1989, J. SIEMONS (ed)
142 The geometry of jet bundles, D.J. SAUNDERS
143 The ergodic theory of discrete groups, PETER J. NICHOLLS
144 Introduction to uniform spaces, I.M. JAMES
145 Homological questions in local algebra, JAN R. STROOKER
146 Cohen-Macaulay modules over Cohen-Macaulay rings, Y. YOSHINO
147 Continuous and discrete modules, S.H. MOHAMED & B.J. MÜLLER
148 Helices and vector bundles, A.N. RUDAKOV et al
149 Solitons, nonlinear evolution equations and inverse scattering, M.A. ABLOWITZ & P.A. CLARKSON
150 Geometry of low-dimensional manifolds 1, S. DONALDSON & C.B. THOMAS (eds)
151 Geometry of low-dimensional manifolds 2, S. DONALDSON & C.B. THOMAS (eds)
152 Oligomorphic permutation groups, P. CAMERON
153 L-functions and arithmetic, J. COATES & M.J. TAYLOR (eds)
154 Number theory and cryptography, J. LOXTON (ed)
155 Classification theories of polarized varieties, TAKAO FUJITA
156 Twistors in mathematics and physics, T.N. BAILEY & R.J. BASTON (eds)
158 Geometry of Banach spaces, P.F.X. MÜLLER & W. SCHACHERMAYER (eds)
159 Groups St Andrews 1989 Volume 1, C.M. CAMPBELL & E.F. ROBERTSON (eds)
160 Groups St Andrews 1989 Volume 2, C.M. CAMPBELL & E.F. ROBERTSON (eds)

London Mathematical Society Lecture Note Series. 159

Groups St Andrews 1989
Volume 1

Edited by
C.M. Campbell
University of St Andrews
and
E.F. Robertson
University of St Andrews

The right of the University of Cambridge to print and sell all manner of books was granted by Henry VIII in 1534. The University has printed and published continuously since 1584.

CAMBRIDGE UNIVERSITY PRESS
Cambridge
New York Port Chester Melbourne Sydney

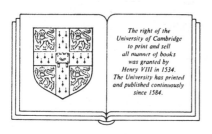

CAMBRIDGE UNIVERSITY PRESS
Cambridge, New York, Melbourne, Madrid, Cape Town, Singapore,
São Paulo, Delhi, Dubai, Tokyo

Cambridge University Press
The Edinburgh Building, Cambridge CB2 8RU, UK

Published in the United States of America by Cambridge University Press, New York

www.cambridge.org
Information on this title: www.cambridge.org/9780521398497

First published 1991

A catalogue record for this publication is available from the British Library

ISBN 978-0-521-39849-7 Paperback

Transferred to digital printing 2010

CONTENTS

VOLUME I

CONTENTS OF VOLUME II

PREFACE

This is the first of two volumes of the Proceedings of Groups - St Andrews 1989. There is a full contents of both volumes and those papers written by authors with the name of the first author lying in the range A-G. Contained in this part are the papers of two of the main speakers namely J A Green and N D Gupta, and we would like especially to thank them for their contributions both to the conference and to this volume.

INTRODUCTION

An international conference 'Groups - St Andrews 1989' was held in the Mathematical Institute, University of St Andrews, Scotland during the period 29 July to 12 August 1989. A total of 293 people from 37 different countries registered for the conference. The initial planning for the conference began in July 1986 and in the summer of 1987 invitations were given to Professor J A Green (Warwick), Professor N D Gupta (Manitoba), Professor O H Kegel (Freiburg), Professor A Yu Ol'shanskii (Moscow) and Professor J G Thompson (Cambridge). They all accepted our invitation and gave courses at the conference of three or four lectures. We were particularly pleased that Professor Ol'shanskii was able to make his first visit to the West. The above courses formed the main part of the first week of the conference. All the above speakers have contributed articles based on these courses to the Proceedings.

In the second week of the conference there were fourteen one-hour invited survey lectures and a CAYLEY workshop with four main lectures. In addition there was a full programme of research seminars. The remaining articles in the two parts of the Proceedings arise from these invited lectures and research seminars.

The two volumes of the Proceedings of Groups - St Andrews 1989 are similar in style to 'Groups - St Andrews 1981' and 'Proceedings of Groups - St Andrews 1985' both published by Cambridge University Press in the London Mathematical Society Lecture Notes Series. Rather than attempt to divide the two parts by an inevitably imprecise division by subject area we have divided the two parts by author name, the first part consisting of papers with author names beginning A-G and the second part author names beginning H-V.

A feature of these volumes is the number of surveys written by leading researchers, in the wide range of group theory covered. From the papers an extensive list of references may be built up and some of the diversity of group theory appreciated.

The computing aspect of group theory was catered for in several ways. As mentioned above there was a series of lectures on CAYLEY and a CAYLEY workshop; there was also a lecture on GAP. Participants were also given access to the Microlaboratory equipped with MAC's, to the University of St Andrews VAX's and to a SUN 3/260 running a wide range of software including CAYLEY, GAP, SPAS and SOGOS. All these facilities were well used by the conference participants.

Groups - St Andrews 1989 received financial support from the Edinburgh Mathematical Society, the London Mathematical Society and the British Council. We gratefully acknowledge this financial support. We would also like to thank Cambridge University Press and the London Mathematical Society for their help with publishing.

We would like to express our thanks to all our colleagues who helped in the running of the conference and in particular Patricia Heggie, John O'Connor and Trevor Walker. We would also like to thank our wives for their help and forbearance. We would like to thank Shiela Wilson for so willingly undertaking the daunting task of typing the two volumes of the Proceedings, having already typed the Proceedings of the previous two St Andrews conferences.

Our final thanks go to those authors who have contributed articles to these volumes. We have edited these articles to produce some uniformity without, we hope, destroying individual styles. For any inconsistency in, and errors introduced by, our editing we take full responsibility.

Colin M Campbell

Edmund F Robertson

TRIPLY FACTORIZED GROUPS

BERNHARD AMBERG

Universität Mainz, D-6500 Mainz, West Germany

Let the group $G = AB$ be the product of two subgroups A and B. If N is a normal subgroup of G, then it is easy to see that the factorizer $X(N) = AN \cap BN$ has the triple factorization

$$X(N) = N(A \cap BN) = N(B \cap AN) = (A \cap BN)(B \cap AN)$$

(see for instance [1]). Therefore it is of interest to consider triply factorized groups of the form

$$G = AB = AK = BK \text{ where K is a normal subgroup of G.}$$

If the group theoretical property $\mathfrak{X}$ is inherited by epimorphic images and extensions, then clearly the triply factorized group G is an $\mathfrak{X}$-group whenever A, B and K are $\mathfrak{X}$-groups. This will usually fail to be the case if $\mathfrak{X}$ is not extension inherited and is for instance some nilpotency or supersolubility requirement. This is most strikingly seen from the following examples. (A survey on results about general infinite factorized groups can be found in [3].)

1. Some examples

In [23] Sysak has given examples of triply factorized groups $G = AB = AK = BK$ where A, B and K are abelian and K is normal in G. Among these there are some which are not nilpotent in any reasonable sense.

1.1 General Construction. Let R be a radical ring and let A be the set R with operation

$$r \circ s = r + s + rs \text{ for every } r, s \in R.$$

Then A is a group which operates on the additive group $M = R^+$ of R via

$$m^r = m \circ r - r = m + mr \text{ for every } m \in M \text{ and } r \in A.$$

Consider the semi-direct product

$$G = A \ltimes M = \{(r,m) \mid r \in A, m \in R\}.$$

Define the following subgroups of G:

$$\{(r,0) \mid r \in A\},$$

$$\{(0,m) \mid m \in R\} \subseteq R^+,$$

$$\{(r,r) \mid r \in A\}.$$

If we identify these with A, M and B respectively, then the following holds:

$$1 = B \cap M = A \cap M = A \cap B.$$

Multiplication in G is given by

$$(r,s)(r',s') = (r + r' + rr',\ s + s' + sr').$$

In particular we have

$$(r,r)(r',0) = (r + r' + rr',\ r + rr') = (0,m) \in M,$$

so that $r + r' + rr' = 0$ and $r' = -m$. This shows that $M \subseteq BA$, and similarly $A \subseteq BM$. Hence $G = AM \subseteq BM \subseteq BA$. We have shown:

$$G = AM = BM = AB,\ A \cap M = B \cap M = A \cap B = 1,\ A, B \text{ and } M \text{ abelian.}$$

Using the above multiplication rule for the semidirect product we compute

$$(0,-s)(r,0)(0,s) = (0,-s)(r,s) = (r,-sr) = (r,0)(0,-sr),$$

so that

$$(0,-sr) = [(r,0),\ (0,s)].$$

We suppose now that the radical ring R satisfies the following additional requirement:

For every $r \neq 0$ there exists an element s in R with $sr \neq 0$. (*)

Assume there exists an element (r,0) in the core A_G of A with $r \neq 0$. Then we have that

$$1 \neq (0,-sr) = [(r,0),\ (0,s)] \in A_G \subseteq A.$$

This contradiction shows $A_G = 1$, and similarly $B_G = 1$. Then also $A \cap Z(G)$ and $B \cap Z(G)$ are trivial, so that it follows from the following lemma that $Z(G) = 1$. In particular G is not hypercentral.

Lemma. *If the group* G = AB *is the product of two abelian subgroups* A *and* B, *then* $Z(G) = (A \cap Z(G))(B \cap Z(G))$.

To prove this let $c = ab^{-1}$ with $a \in A$ and $b \in B$ be a non-trivial element of Z(G). Let $s = a^*b^*$ be an arbitrary element of G with $a^* \in A$ and $b^* \in B$. Then

$$[a,s] = [a, a^*b^*] = [a,b^*][a,a^*]^{b^*} = [a,b^*] = [cb,b^*]$$
$$= [c,b^*]^b[b,b^*] = [c,b^*]^b = 1.$$

Therefore $a \neq 1$ is contained in Z(G) and similarly also b is in Z(G). This proves the lemma.

1.2 Example. Let $p \neq 2$ be a prime, and let R be the ring of all rational numbers of the form u/v where u and v are integers with $v \neq 0$, p divides u, but p does not divide v. Then R is a radical ring and clearly it also satisfies condition (*). Clearly $M = R^+$ as a subgroup of $\mathbf{Q}^+$ is torsion-free of Prüfer rank 1. The groups A and B are free abelian of countable infinite rank. Hence also the group G is torsion-free. It can be shown that G is not locally nilpotent and not even locally polycyclic.

2. Finite groups

An epimorphism inherited class of finite groups $\mathfrak{F}$ is a saturated formation if for every finite group G the following holds: (i) $G/(N \cap M)$ is an $\mathfrak{F}$-group whenever G/N and G/M are $\mathfrak{F}$-groups and (ii) G is an $\mathfrak{F}$-group whenever the Frattini factor group G/Frat G is an $\mathfrak{F}$-group. Examples for saturated formations are the classes of finite nilpotent and finite supersoluble groups.

Theorem 2.1. *Let* $\mathfrak{F}$ *be a saturated formation of finite groups, and let the group* G = AB = AK = BK *be the product of three subgroups* A, B *and* K, *where* K *is normal in* G. *If* A *and* B *are* $\mathfrak{F}$*-groups and* K *is nilpotent, then* G *is an* $\mathfrak{F}$*-group.*

Proof. Assume that Theorem 2.1 is false, and let G = AB = AK = BK be a counterexample of minimal order. Then every proper epimorphic image of G is an $\mathfrak{F}$-group. If G has two different minimal normal subgroups N and M, then $G \simeq G/(N \cap M)$ is an $\mathfrak{F}$-group. This contradiction shows that G has exactly one minimal normal subgroup M.

Since M is contained in the nilpotent normal subgroup K of G, M is abelian. The Frattini subgroup Frat G is trivial, since $\mathfrak{F}$ is saturated. Hence M has a complement L in G, so that G = ML and $M \cap L = 1$. Assume that $M \subset C_G(M)$. then $C_G(M) \cap L$ is normal in ML = G and hence must be trivial. Then

$|C_G(M)L| > |ML| = |G|$.

This contradiction shows that $M = C_G(M)$. Then also $M = K$ since K is nilpotent.

Let U be a maximal subgroup of G containing A. Then $U = U \cap AK = A(K \cap U)$, where $K \cap U$ is a proper G-invariant subgroup of K. Therefore $K \cap U = 1$ and $A = U$ is a maximal subgroup of G. Similarly B is a maximal subgroup of G. - Since K is contained in every non-trivial normal subgroup of G, it follows that A and B are $\mathfrak{F}$-maximal in every epimorphic image of G, so that they are $\mathfrak{F}$-projectors of G. But it is well-known that the $\mathfrak{F}$-projectors of a finite group are conjugate (see P Schmid [22]), so that $A = B = G$ is an $\mathfrak{F}$-group. This contradiction proves the theorem.

Corollary 2.2. *Let the finite group* $G = AB = AK = BK$ *be the product of three subgroups* A, B *and* K *where* K *is nilpotent and normal in* G. *If* A *and* B *are nilpotent (supersoluble), then* G *is nilpotent (supersoluble).*

2.3 Remark. There exist finite groups which are not supersoluble, but all their proper subgroups are supersoluble and whose orders contain exactly three primes. These have a triple factorization $G = AB = AK = BK$ with three supersoluble subgroups A, B and K, where K normal in G. This shows that in 2.1 and 2.2 the normal subgroup K has to be nilpotent.

2.4 Problem. Does Theorem 2.1 still hold when the subgroup K of

$$G = AB = AK = BK$$

is no longer normal in G? (Kegel has shown in [16] that Corollary 2.2 holds when the subgroup K is not normal in G.)

3. Locally finite groups

A locally finite group $G = AB = AK = BK$ which is the product of three abelian subgroups A, B and K, where K is normal in G, need not be nilpotent, as the following example shows.

3.1 Example. For each odd prime p let G_p be a metacyclic p-group of class $\geq p$ which has a triple factorization

$$G_p = A_pB_p = A_pK_p = B_pK_p,$$

where A_p, B_p and K_p are cyclic and K_p is normal in G_p (see [24, Example 1]). The direct product $G = \mathrm{Dr}_p\, G_p$ can be written as

$$G = AB = AK = BK,$$

where $A = Dr_p A_p$, $B = Dr_p B_p$ and $K = Dr_p K_p$. It is easy to see that G satisfies the required conditions, is not nilpotent and has Prüfer rank 2.

3.2 Remark. In [15] Holt and Howlett give an example of a metabelian p-group $G = AB = AK = BK$ where A and B are elementary abelian subgroups of G and K is an abelian normal subgroup of G; G is locally nilpotent, but not hypercentral.

The following two results are proved in Amberg [2].

Theorem 3.3. *Let the locally finite group* $G = AB = AK = BK$ *be the product of two hypercentral subgroups* A *and* B *and a locally nilpotent normal subgroup* K *of* G. *Then* G *is locally nilpotent.*

The proof of this result relies heavily on the fact that the hypercentral subgroups $A \neq 1$ and $B \neq 1$ have non-trivial centres.

Theorem 3.4. *Let the locally finite group* $G = AB = AK = BK$ *be the product of three locally nilpotent subgroups* A, B *and* K, *where* K *is normal in* G. *If for every prime* p *the maximal* p*-subgroups of* G *are conjugate, then* G *is locally nilpotent.*

3.5 Problem. Let the locally finite group $G = AB = AK = BK$ be the product of three subgroups A, B and K, where K is locally nilpotent and normal in G. If A and B are locally nilpotent (locally supersoluble), under which conditions is G then locally nilpotent (locally supersoluble)?

To mention some further positive answers to these questions we recall the definition of a class of locally finite-soluble groups that was introduced by Gardiner, Hartley and Tomkinson in [13].

Let $\mathfrak{U}$ be the largest subgroup closed class of locally finite groups with the following properties:

(i) Every $\mathfrak{U}$-group has a finite (subnormal) series with locally nilpotent factors,

(ii) For every set of primes π the maximal π-subgroups of G are conjugate.

Now let $\mathfrak{K}$ be a subgroup and image closed subclass of $\mathfrak{U}$.

A $\mathfrak{K}$-formation $\mathfrak{F}$ is a class of $\mathfrak{K}$-groups which is closed under the forming of epimorphic images and which is *residual with respect to* $\mathfrak{K}$, i.e. if

$$G \in \mathfrak{K},\ G/N_\alpha \in \mathfrak{F} \text{ for every } \alpha \text{ in the index set I, then } G/\bigcap_\alpha N_\alpha \in \mathfrak{F}.$$

In the paper of Gardiner, Hartley and Tomkinson, *saturated* $\mathfrak{K}$-formations are then defined which coincide for finite groups with the usual definitions. The following generalization of a theorem of Carter and Hawkes can be proved:

Let $\mathfrak{F}$ be a saturated $\mathfrak{K}$-formation. If the $\mathfrak{F}$-residual D *of the $\mathfrak{K}$-group* G *is abelian, then* D *has complements in* G *and all these complements are conjugate in* G.

Using this Amberg and Halbritter have proved the following result (unpublished).

Theorem 3.6. *Let $\mathfrak{F}$ be a saturated $\mathfrak{K}$-formation. Let the $\mathfrak{K}$-group*

$$G = AB = AK = BK$$

be the product of three subgroups A, B *and* K, *where* K *is nilpotent and normal in* G. *If* A *and* B *are $\mathfrak{F}$-subgroups, then* G *is an $\mathfrak{F}$-group.*

The proof of this theorem is by induction on the nilpotency class of K, but it seems likely that the result also holds when K is only locally nilpotent.

4. Mixed groups

A soluble group G is a *minimax group* if it has a finite series whose factors are finite or infinite cyclic or quasicyclic of type p^∞; the number of infinite cyclic factors in such a series is the *torsion-free rank* $r_0(G)$ of G, the number of factors of type p^∞ for the prime p is the p^∞*-rank* $m_p(G)$ of G and the *minimax rank* of G is

$$m(G) = r_0(G) + \sum_p m_p(G).$$

The following general theorem contains results that have been provided in a number of papers by Amberg, Franciosi and de Giovanni (see [5], [6], [7], [9]; see also Halbritter [14]).

Theorem 4.1. *Let the group* G = AB = AK = BK *be the product of two subgroups* A *and* B *and a normal minimax subgroup* K *of* G.

(a) *If* A, B *and* K *satisfy some nilpotency requirement $\mathfrak{N}$, then* G *satisfies the same nilpotency requirement $\mathfrak{N}$. Here $\mathfrak{N}$ can for instance be chosen to be any of the following group classes:*

(i) *the classes of nilpotent, hypercentral or locally nilpotent groups,*

(ii) *the classes of* FC*-nilpotent,* FC*-hypercentral groups and locally* FC*-nilpotent groups,*

(iii) *the classes of nilpotent-by-$\mathfrak{X}$, hypercentral-by-$\mathfrak{X}$ and (locally nilpotent)-by-$\mathfrak{X}$ groups, where $\mathfrak{X}$ is the class of finite, periodic, polycyclic, Chernikov or minimax groups,*

(iv) *the class of finite-by-nilpotent groups,*

(v) *the class of locally polycyclic groups.*

(b) *Let* K *be hypercentral. If* A *and* B *are supersoluble (hypercyclic resp. locally supersoluble), then* G *is supersoluble (hypercyclic resp. locally supersoluble).*

Remarks on the proof of Theorem 4.1. (a) In the proof of this general theorem each of the statements has to be checked individually. The proofs are by way of contradiction, inducting on the minimax rank m(K).

(b) The case when K is nilpotent can be reduced to the case when K is abelian by using so called *'theorems of Hall's type'*:

If K *is a nilpotent normal subgroup of the group* G *such that* G/K' *is an* $\mathfrak{N}$*-group, then* G *is an* $\mathfrak{N}$*-group.*

For the class $\mathfrak{N}$ of nilpotent groups this is a well-known result of P Hall (see [19, Part 1, Theorem 2.27, p.56]). For most of the other group classes $\mathfrak{N}$ in Theorem 4.1 the above statement can also be proved.

(c) When K is a torsion-free abelian minimax group, some facts about the automorphism group Aut K are used. In particular a theorem of Baer depending on Dirichlet's Unit theorem of Algebraic Number Theory says that in this case *soluble subgroups of* Aut K *are minimax groups.*

(d) Essential use is also made of cohomological results such as the following theorem of D Robinson in [20]. As usual, $H^i(Q,M)$ denotes the i-th cohomology group of the group Q with coefficients in the Q-module M.

Let Q *be a locally nilpotent group and* M *a* Q*-module such that* $Q/C_Q(M)$ *is hypercentral and* $H^0(Q,M) = 0$. *If* M *is an artinian* Q*-module, then* $H^n(Q,M) = 0$ *for every non-negative integer* n.

These cohomological results can be used to show that the two supplements A and B of K in G are in fact complements and that they are conjugate which easily leads to a contradiction as in the proof of Theorem 2.1.

Three lemmas

The proof of the first statement of Theorem 4.1 that G is nilpotent whenever A, B and K are nilpotent, is much simpler and does not require cohomological results or facts depending on algebraic number theory. The following *'extension lemma'* gives a useful criterion for a group with a nilpotent triple factorization to be nilpotent (see Robinson [21]).

Lemma 4.2. *Let the group* G = AB = AK = BK *be the product of three nilpotent subgroups* A, B *and* K, *where* K *is normal in* G, *and assume that the Baer radical of* G *is nilpotent. If there exists a normal subgroup* N *of* G *such that the factorizer* X(N) *of* N *in* G *and the factor group* G/N *are nilpotent, then* G *is nilpotent.*

Proof. Clearly K is contained in the Baer radical R of G. Now X(N) is subnormal in G since it contains N; therefore $X(N) \subseteq R$. Since $G = AK$,

$$[A \cap BN, {}_rG] \subseteq [A \cap BN, {}_rA]\, R' = R'$$

for a sufficiently large integer r. Therefore $(A \cap BN)R'/R'$ is contained in some term of the upper central series of G/R' with finite ordinal type. Of course a similar statement is true of $(B \cap AN)R'/R'$, so it follows that NR'/R' is contained in some term with finite ordinal type of the upper central series of G/R'. Consequently G/R' is nilpotent. By hypothesis R is nilpotent and a well-known theorem of P Hall shows that G is nilpotent.

To apply the 'extension lemma' we need to know that the Baer radical of G is nilpotent. This is ensured by the next lemma.

Lemma 4.3. *Let* N *be a normal subgroup with finite Prüfer rank* r *of the Baer group* G.

(a) *If* N *is a radicable abelian* p*-group, then* N *is contained in the* r*-th term* $Z_r(G)$ *of the upper central series of* G.

(b) *If the torsion subgroup of* N *is a Chernikov group and the factor group* G/N *is nilpotent, then* G *is nilpotent.*

Proof. (a) If H is a finitely generated subgroup of G, then H is a nilpotent subnormal subgroup of G. Write

$$N_i = [N, \underbrace{H, \ldots, H}_{i}]$$

for every positive integer i. Then $N_t = 1$ for some t. Since every N_i is radicable, it follows that N_i is a direct factor of N_{i-1} for all $i \leq t$. Hence, $t \leq r$ and thus $N_r = 1$. Therefore also $[N, \underbrace{G, \ldots, G}_{r}] = 1$ and so $N \subseteq Z_r(G)$.

(b) Since the torsion subgroup T of N is a Chernikov group, its finite residual J is a radicable abelian torsion group with finite Prüfer rank, and T/J is finite. Clearly N/T is a torsion-free nilpotent normal subgroup of G/T with finite Prüfer rank and so $N/T \subseteq Z_s(G/T)$ for some positive integer s (see [17, Part 2, p.35]) and G/T is nilpotent. Thus G/J is finite-by-nilpotent and hence nilpotent. By (a) we have that $J \subseteq Z_r(G)$, so that G is nilpotent.

The last lemma is elementary, but very useful (see Amberg [4]).

Lemma 4.4. *Let the group* $G = AB = AK = BK$ *be the product of two subgroups* A *and* B *and an abelian normal subgroup* $K \neq 1$ *of* G *with*

$$A \cap K = B \cap K = 1.$$

If $a = bx \in Z(A)$ *with* $b \in B$ *and* $x \in K$, *and if* $N \neq 1$ *is a normal subgroup of* G *contained in* K *and containing* x, *then* $[N,a]$ *is a normal subgroup of* G *with* $[N,a] \subset N$.

Proof. Clearly $Z(A)K = Z(B)K$ and hence $b \in Z(B)$. Since N is an abelian normal subgroup of G, $[N,a]$ is the set of all $[n,a]$ with $n \in N$. Then $[N,a]$ is a normal subgroup of $G = AK$, as $a \in Z(A)$.

If $a \in Z(B)$, then $a \in Z(G)$ and therefore $[N,a] = 1 \subset N$. Hence we may assume that $a \notin Z(B)$, so that $x \neq 1$. Assume that $N = [N,a]$. There exists an element $n = a^*b^* \neq 1$ in N with $1 \neq x = [n,a]$ with $a^* \in A$ and $b^* \in B$. It follows that

$$[a^*b^*,a] = b^{*-1}a^{*-1}a^{-1}a^*b^*a = b^{*-1}a^{-1}b^*a$$

and therefore

$$a = bx = bb^{*-1}a^{-1}b^*a$$

and

$$b^*b^{-1}b^{*-1} = a^{-1} \in B.$$

Since $a \in Z(A)$ and $b \in Z(B)$ it follows that $a = b \in Z(B)$. This contradiction proves the lemma.

Proof of the 'nilpotent statement' of Theorem 4.1.

Let $G = AB = AK = BK$, *where* A, B *and* K *are nilpotent subgroups of* G *and* K *is a minimax group. Then* G *is nilpotent.*

Assume that this statement is false, and choose among the counterexamples with K of minimal minimax rank a group G for which the sum of the nilpotency classes of A and B is also minimal. By a well-known result of P Hall we may assume that K is abelian (see [19, Part 1, p.56]). If G is finite-by-nilpotent, $|G:Z_n(G)|$ is finite for some non-negative integer n by another theorem of P Hall (see [19, Part 1, p.117]). Since the nilpotent statement holds for finite groups by Theorem 2.1 or Corollary 2.2 it follows that G is nilpotent. This contradiction shows that G is not finite-by-nilpotent.

(i) The case: K is a torsion group

In this case K is a Chernikov group. There exists a finite G-invariant subgroup E of K such that K/E is radicable. If G/E is nilpotent, then G is finite-by-nilpotent, a contradiction. Therefore we may assume that K is radicable.

Let H be an infinite G-invariant subgroup of K. If $H \subset K$, then the factor group G/H and the factorizer X(H) of H in G are nilpotent. By Lemma 4.3(b) the Baer radical of G is nilpotent, so that G is nilpotent by Lemma 4.2. This contradiction shows that every proper G-invariant subgroup of K is finite.

Clearly the normal subgroups $A \cap K$ and $B \cap K$ of G are properly contained in K, so that $C = (A \cap K)(B \cap K)$ is a finite normal subgroup of G and so G/C is not nilpotent. Therefore we may assume that $A \cap K = B \cap K = 1$. For every a in Z(A), the group [K,a] is a normal subgroup of G which is properly contained in K; see Lemma 4.4. Since K is radicable, we have that [K,a] = 1. This shows that $Z(A) \subseteq Z(G)$, so that G/Z(G) is nilpotent. This contradiction proves that G is nilpotent when K is a torsion group.

(ii) The general case

The factorizer X(T) of the torsion subgroup T of K is nilpotent by case (i). By Lemma 4.3(b) the Baer radical of G is nilpotent, so that G/T is not nilpotent by Lemma 4.2. Hence we may assume that K is torsion-free.

Then K/K^p is finite for every prime p, so that by case (i) G/K^p is nilpotent. It follows that $[K, {}_rG] \subseteq K^p$ where r is the rank of K. Since K is a minimax group we have $\bigcap_p K^p = 1$. Consequently $[K, {}_rG] = 1$. Hence G is nilpotent and this contradiction proves the nilpotent statement of Theorem 4.1.

5. An Application

The 'nilpotent statement' of Theorem 4.1 has the following consequence, which generalizes results in [1] for polycyclic groups and of Pennington in [17] for finite groups.

Theorem 5.1. *If the soluble minimax group* G = AB *is the product of two nilpotent subgroups* A *and* B, *then each term of the ascending Fitting series of* G *is factorized; in particular the Fitting subgroup of* G *is factorized.*

Proof. Consider the ascending Fitting series of G

$$1 = F_0 \leq F_1 \leq F_2 \leq \ldots F_n = G,$$

where F_i is normal in F_{i+1} and F_{i+1}/F_i is the Fitting subgroup of G/F_i; in particular $F_1 = \mathrm{Fitt}(G)$. If X_i is the factorizer of F_i in G for $i \leq n$, then the subgroup X_{i+1}/F_i is the factorizer of F_{i+1}/F_i in G/F_i. Hence X_{i+1}/F_i is nilpotent by Theorem 4.1. Since $F_i \leq X_i \leq X_{i+1}$, the subgroup X_i is subnormal in X_{i+1} for every $i < n$. It follows that $X_1 = X(F_1) = X(\mathrm{Fitt}(G))$ is a nilpotent subnormal subgroup of G, so that it is contained in the Baer radical which coincides with the Fitting subgroup of G (see [19, Part 2, p.38]). This proves the theorem.

6. Soluble groups with finite abelian section rank

The examples of Sysak show that Theorem 4.1 does not hold for soluble groups of finite Prüfer rank. However for soluble groups with finite abelian section rank there is a corresponding result. Recall that a group has *finite abelian section rank* if every elementary abelian p-section is finite for every prime p.

Theorem 6.1. *Let* $G = AB = AK = BK$ *be a soluble group with finite abelian section rank, where* A *and* B *are subgroups and* K *is a normal subgroup of* G.

(a) *If* A, B *and* K *satisfy some nilpotency requirement* $\mathfrak{N}$, *then* G *satisfies the same nilpotency requirement* $\mathfrak{N}$. *Here* $\mathfrak{N}$ *can be any of the group classes of Theorem 6.1 except the classes of nilpotent and finite-by-nilpotent groups.*

(b) *If* K *is hypercentral and* A *and* B *are supersoluble (hypercyclic or locally supersoluble respectively), then* G *is supersoluble (hypercyclic or locally supersoluble respectively).*

That there is no corresponding result for the classes of nilpotent and of finite-by-nilpotent groups can be seen from Example 3.1.

Using Theorem 6.1 for the class of locally nilpotent groups the following corollary can be proved in a similar way as Theorem 5.1.

Corollary 6.2. *Let the soluble group* $G = AB$ *with finite abelian section rank be the product of two subgroups* A *and* B.

(a) *If* A *and* B *are locally nilpotent, then the Hirsch-Plotkin radical* $R = R(G)$ *is factorized.*

(b) *If* A *and* B *are (locally nilpotent)-by-$\mathfrak{X}$ - where $\mathfrak{X}$ is as in Theorem 4.1 - then the factor group* $X(R)/R$ *is an $\mathfrak{X}$-group.*

6.3 Problem. Does Corollary 6.2 still hold for a soluble group $G = AB$ with finite torsion-free rank? (Recall that a group has *finite torsion-free rank* when it has a finite (subnormal) series in which each non-periodic factor is infinite cyclic.)

7. Trifactorized groups

In [16] Kegel has shown that a 'trifactorized' finite group G = AB = AC = BC where A, B and C are nilpotent subgroups, is likewise nilpotent (for another proof of this see Pennington [18]). Some extensions of this theorem to almost soluble minimax groups can be found in Amberg, Franciosi and de Giovanni [18] and Amberg and Halbritter [10].

Theorem 7.1. *Let the soluble minimax group* G = AB = AC = BC *be the product of three subgroups* A, B *and* C.

(a) *If* A, B *and* C *are nilpotent, then* G *is nilpotent.*

(b) *If* A, B *and* C *are finite-by-nilpotent, then* G *is finite-by-nilpotent.*

(c) *If* A *and* B *are nilpotent and* C *is locally supersoluble, then* G *is locally supersoluble and hence hypercyclic.*

(d) *If* A *and* B *are nilpotent and* C *is locally* FC-*nilpotent, then* G *is locally* FC-*nilpotent and hence* FC-*nilpotent.*

Problems. (a) Find other statements like those in Theorem 7.1 for soluble minimax groups.

(b) Can 'soluble minimax' in Theorem 7.1 be replaced by 'soluble and finitely generated'. (For instance the first statement in Theorem 7.1 still holds since it is well-known that a finitely generated soluble group which is not nilpotent has a finite epimorphic image which is not nilpotent).

(c) Let $\mathfrak{X}$ be a class of groups which is closed under the forming of subgroups, epimorphic images and extensions. Let the group G = AB = AC = BC be the product of three $\mathfrak{X}$-subgroups A, B and C. Under which conditions is G an $\mathfrak{X}$-group?

References

1. B Amberg, Artinian and noetherian factorized groups, *Rend. Sem. Mat. Univ. Padova* **55** (1976), 105-122.

2. B Amberg, Lokal endlich-auflösbare Produkte von zwei hyperzentralen Gruppen, *Arch. Math. (Basel)* **35** (1980), 228-238.

3. B Amberg, *Infinite factorized groups* (Groups - Korea 1988, Proceedings, Pusan 1988, Lecture Notes in Mathematics **1398**, Springer-Verlag).

4. B Amberg, Produkte von Gruppen mit endlichem torsionsfreiem Rang, *Arch. Math. (Basel)* **45** (1985), 398-406.

5. B Amberg, S Franciosi & F de Giovanni, Groups with an FC-nilpotent triple factorization, *Ricerche Mat.* **36** (1987), 103-114.

6. B Amberg, S Franciosi & F de Giovanni, Groups with a nilpotent triple factorization, *Bull. Austral. Math. Soc.* **37** (1988), 69-79.

7. B Amberg, S Franciosi & F de Giovanni, Groups with a supersoluble triple factorization, *J. Algebra* **117** (1988), 136-148.

8. B Amberg, S Franciosi & F de Giovanni, On trifactorized soluble minimax groups, *Arch. Math. (Basel)* **51** (1988), 13-19.

9. B Amberg, S Franciosi & F de Giovanni, Triply factorized groups, *Comm. Algebra*, to appear.

10. B Amberg & N Halbritter, On groups with a triple factrorization, *Ricerche Mat.*, to appear.

11. B Amberg & D J S Robinson, Soluble groups which are products of nilpotent minimax groups, *Arch. Math. (Basel)* **42** (1984), 385-390.

12. S Franciosi & F de Giovanni, On products of locally polycyclic groups, *Arch. Math.(Basel)* , to appear.

13. A D Gardiner, B Hartley & M J Tomkinson, Saturated formations and Sylow structure in locally finite groups, *J. Algebra* **17** (1971), 177-211.

14. N Halbritter, Groups with a nilpotent-by-finite triple factorization, *Arch. Math. (Basel)* **51** (1988), 393-400.

15. D F Holt & R B Howlett, On groups which are the product of two abelian groups, *J. London Math. Soc.* *29* (1984), 453-461.

16. O H Kegel, Zur Struktur mehrfach faktorisierter endlicher Gruppen, *Math. Z.* **87** (1965), 42-48.

17. E Pennington, On products of finite nilpotent groups, *Math. Z.* **134** (1973), 81-83.

18. E Pennington, Trifactorisable groups, *Bull. Austral. Math. Soc.* **8** (1973), 461-469.

19. D J S Robinson, Finiteness conditions and generalized soluble groups, (Springer-Verlag, Berlin, 1972).

20. D J S Robinson, The vanishing of certain homology and cohomology groups, *J. Pure Appl. Algebra* **7** (1976), 145-167.

21. D J S Robinson, Soluble products of nilpotent groups, *J. Algebra* **98** (1986), 183-196.

22. P Schmidt, Lokale Formationen endlicher Gruppen, *Math. Z.* **137** (1974), 31-48.

23. Y P Sysak, Products of infinite groups, *Preprint 82.53, Akad. Nauk. Ukrain. Mat. Kiev* (1982), 1-36.

24. D I Zaicev, Nilpotent approximations of metabelian groups, *Algebra and Logic* **20** (1981), 413-423.

AN INTRODUCTION TO A CLASS OF TWO RELATOR GROUPS

IRIS LEE ANSHEL

Columbia University, New York, NY 10027, USA

1. Introduction

The classical Freiheitssatz of W Magnus for groups with a single defining relator states that if the group G is given by the presentation

$$G = \langle t, x_1, \dots, x_n \mid r \rangle \tag{1.1}$$

where the relator r is reduced and cyclically reduced and involves the generator t, then the map induced by inclusion of generators on the free group based on $\{x_1, \dots, x_n\}$

$$\langle x_1, \dots, x_n \mid - \rangle \to G$$

is monic. The Magnus approach to the one relator groups has led to a powerful and rich theory (for detailed accounts the reader is refered to the texts [M-K-S], [L-S] and the survey articles [H], [B]), and we are thus motivated to consider the two relator group

$$G = \langle s, t, A \mid p, q \rangle \tag{1.2}$$

where $A = \{a_1, \dots, a_n\}$, and the relators $p, q \in \langle s, t, A \mid - \rangle \sim \langle A \mid - \rangle$. The behaviour of groups with two defining relators lacks the rigidity of the one relator case: in fact, as we shall see, the subgroup of G generated by A is not in general free (see (2.1)). Bearing such examples in mind we are led to a notion of the defining relators p,q being independent relative to the generators {s,t} (see Section 2). It is with this constraint in place that a Freiheitssatz for the group (1.2) may be proved: it states that if the relators p,q are independent relative to the generators {s,t} then the homomorphism induced by inclusion of generators

$$\langle A \mid - \rangle \to G$$

is monic. The methods used to analyze our two relator groups involve the Bass-Serre theory of graphs of groups (see [S]) in a manner which parallels the use of generalized free products in Magnus' proof (see [MKS]). We can, using our methods, derive applications of our theorem which are analogous to those derived from the one relator Freiheitssatz. Though no details are given here we do state our results and refer the reader to [A] for details.

2. Preliminaries and statement of theorems

The complications which may arise in a two relator group appear even in the simplest of examples. Consider, for example, the group

$$G_1 = < a,s,t \mid a^{\alpha}wa^{\beta}w^{-1}, a^{\gamma}wa^{\delta}w^{-1} > \qquad (2.1)$$

where

$$w \in < s,t \mid - > \sim \{1\}$$

and

$$\alpha, \beta, \gamma, \delta \in \mathbf{Z} \sim \{0\}.$$

Observe that in G_1, the subgroup generated by a is not in general free; in fact

$$a^{\alpha\delta-\beta\gamma} = 1.$$

This identity is derived as follows: from the relators of G_1 we have that,

$$a^{\alpha} = wa^{-\beta}w^{-1} \Rightarrow a^{\alpha\delta} = wa^{-\beta\delta}w^{-1}$$

$$a^{\gamma} = wa^{-\delta}w^{-1} \Rightarrow a^{\alpha\delta} = wa^{-\beta\delta}w^{-1}$$

and thus,

$$a^{\alpha\delta} = a^{\gamma\beta} \qquad \text{i.e.} \quad a^{\alpha\delta-\alpha\gamma} = 1.$$

A somewhat more involved example is given by

$$G_2 = < a,s,t \mid a^{\alpha}w^2a^{\beta}w^{-2}, a^{\gamma}w^3a^{\delta}w^{-3} > \qquad (2.2)$$

where again

$$w \in < s,t \mid - > \sim \{1\}$$

and

$$\alpha,\beta,\gamma,\delta \in \mathbf{Z} \sim \{0\}.$$

Here we may derive the identity

$$a^{\alpha^3\delta^2+\gamma^2\beta^3} = 1.$$

The above examples demonstrate that the subwords of the relators which involve the generators {s,t} must be "independent" in some appropriate sense, and it is this concept which we now proceed to formalize.

Let us being by focusing our attention on the general presentation

$$G = < s,t,A \mid p,q > = F/U$$

where

$$A = \{a_1,...,a_n\}, \quad F = < s,t,A \mid - >$$

and

$$U = < p,q >_F,$$

the normal subgroup of F generated by {p,q}. Let us denote the natural projection from F onto G by

$$z \mapsto z_G = z \bmod U.$$

Following Magnus' model we begin our investigation by considering

$$N = < A,p,q >_F$$

and

$$N_G = N/U = < A_G >_G.$$

Let W be a minimal Schreier coset representative system for F/N ($\cong G/N_G$) (for the existence of such a W, see [MKS, p.93]). Observe that W being a minimal coset representative system insures that

$$W \subseteq < s,t \mid - >:$$

any element w in W which involves an element of A would not have minimal length in the coset it represents.

With our choice of coset representatives $W \subseteq < s,t \mid - >$ now fixed, we have by definition, for each $z \in F$, the decomposition $z = \sigma(z)\overline{z}$ where $\sigma(z) \in N$ and $\overline{z} \in W$. With the notation in place we may now define the key concept in our theorem.

Definition 2.1. The relators p and q are termed independent relative to {s,t} if, for some choice of minimal Schreier coset representative system as above, the following conditions hold.

(a) The relators are given in the form

$$p = x_1c_1x_2c_2 \dots c_k, \qquad q = y_1d_1y_2d_2 \dots d_h$$

where

$$c_i,d_j \in \langle A \mid - \rangle \sim \{1\}, \quad x_i,y_j \in \langle s,t \mid - \rangle \sim \{1\},$$

$\overline{x_i} \neq 1$ for $i = 1,\dots,k$ and $\overline{y_j} \neq 1$ for $j = 1,\dots,h$.

(Remark that $h,k \geq 2$ is implied by this, since $k = 1$ would imply $1 = \overline{x_{(k)}} = \overline{x_1} \neq 1$).

(b) Set

$$P = \{ \overline{x_1}, \overline{x_1x_2}, \dots, \overline{x_1 \dots x_k} = 1\}$$

$$Q = \{ \overline{y_1}, \overline{y_1y_2}, \dots, \overline{y_1 \dots y_h} = 1\}$$

$$R = P \cup Q \subseteq W.$$

(i) $P \not\subseteq Q$, $Q \not\subseteq P$

(ii) $R \cap \overline{R^{-1}} = \{1\}$, and the map $\langle R \sim \{1\} \mid - \rangle \to F/N$ is monic.

We can now state our main theorem.

Theorem 2.2. *Let* $G = \langle s,t,A \mid p,q \rangle$ *be as in (1.2). Assume the relators* p *and* q *are independent relative to* $\{s,t\}$. *Then the natural mapping*

$$\langle A \mid - \rangle \to G$$

is monic.

Observe first that in the group G_1 of (2.1),

$$P = Q = \{1,w\}$$

violating hypothesis (b)(i). Further in the group G_2 of (2.2) we have F/N is free on $\{s,t\}$, and

$$R = \{1,w^2,w^3\}$$

violating hypothesis (b)(ii). Hence our hypothesis immediately eliminates our negative examples.

Obtaining examples of groups where the theroem applies depends mainly on being able to verify hypothesis (b) (ii) in the group G/N_G. The case of G/N_G being free on $\{s,t\}$ is the easiest to compute in. Consider for example the group

$$G_3 = < A,s,t \mid C_1[s,t]C_2[t,s],\ C_3[s^2,t]C_4[t,s^2] > \qquad (2.3)$$

where $C_i \in < A \mid - > \sim \{1\}$ and (in general) $[x,y] = xyx^{-1}y^{-1}$ denotes the commutator. To verify the hypothesis of our theorem in this case notice that indeed G_3/N_{G_3} is free on $\{s,t\}$, and that

$$R \sim \{1\} = \{[s,t], [s^2,t]\}$$

is a subset of a free basis of the commutator subgroup of the free group on $\{s,t\}$.

In the case G/N_G is a one-relator group we may utilize Magnus' theorem in the verification of the hypothesis of our theorem; consider for example the group

$$G_4 = < A,s,t \mid C_1sts^{-1}C_2st^{-1}s^{-1},\ C_3s^3t^2s^{-3}C_4tsts^{-1} > \qquad (2.4)$$

where again $C_i \in < A \mid - > \sim \{1\}$. For a final example consider the group

$$G_5 = < A,s,t \mid C_1xC_2x^{-1}s^3,\ C_3yC_4y^{-1}t^2 > \qquad (2.5)$$

where $C_i \in < A \mid - > \sim \{1\}$ and

$$x = [s,t], \qquad y = [s^2,t].$$

Here

$$G_5/N_{G_5} \cong < s,t \mid s^3, t^2 > \cong PSL_2(\mathbf{Z}).$$

By applying the Reidemeister-Schreier method to $PSL_2(\mathbf{Z})$ we find (see [M-K-S], [L-S]) $R \sim \{1\} = \{x,y\}$ is a subset of a free basis for the commutator subgroup of $PSL_2(\mathbf{Z})$, and we have verified our hypotheses for G_5.

We conclude this section by stating some of the applications of our Freiheitssatz which were alluded to in Section 1.

Theorem 2.3. *Let* $G = < s,t,A \mid p,q >$ *where the relators* p, q *are independent relative to the generators* $\{s,t\}$. *Assume*

$$G/N_G \cong < s,t \mid - >,$$

then the following hold:

(a) G *has solvable word problem.*

(b) *If* card(A) ≥ 2 *then* G *has trivial centre. In the case* card(A) = 1, *if either* card(P) *or* card(Q) ≥ 3 *then* G *has trivial centre (here* P *and* Q *are as in Definition 2.1).*

(c) *The group* G *is torsion free if and only if neither of the relators* p,q *are proper powers.*

(d) *The cohomological dimension of* G *is less than or equal to three if* G *is torsion free.*

The proof of Theorem 2.3 may be found in [A].

3. The proof

Recall that

$$F = < s,t,A \mid - >$$

and

$$G = < s,t,A \mid p,q >$$

where $A = \{a_1,...,a_n\}$. In order to investigate the structure of $< A_G > \leq G$, we begin by obtaining a presentation for the inverse image of $< A_G > \leq G$, $N = < A,p,q >_F$ in F, and for $N_G = < A_G > \leq G$. With this done our plan is to construct a non-trivial homomorphism from $N_G = < A_G > \leq G$ to the fundamental group of a specially constructed graph of groups, and with this accomplished our theorem shall be reduced to verifying that the free group based on A embeds in a vertex group of our graph of groups.

To begin, we choose as before $W \subseteq < s,t \mid - >$ a minimal Schreier coset representative system for F/N (see Section 2), and we have for $z \in F$

$$z = \sigma(z)\overline{z}$$

where $\sigma(z) \in N$, and $\overline{z} \in W$. In our calculations we shall use the following properties of σ. For all $z,z_1,...,z_n \in F$ we have

$$\text{(i)} \quad \sigma(z_1 \ldots z_n) = \sigma(z_1)\sigma(\overline{z_1}\, z_2) \ldots \sigma(\overline{z_1 \ldots z_{n-1}}\, z_n) \qquad (3.1)$$

$$\text{(ii)} \quad \sigma(\overline{z_1}\, z^{-1}) = \sigma(\overline{z_1 z^{-1}}\, z)^{-1}.$$

Applying the Reidemeister-Schreier method (see [MKS] or [L-S]), we find that N is freely generated by the non-trivial elements in the family

$$\sigma(ws), \sigma(wt), \sigma(wa_1), ..., \sigma(wa_n) \qquad (w \in W). \qquad (3.2)$$

Note that

$$\sigma(wc) = wc(\overline{wc})^{-1} = wcw^{-1} \qquad (3.3)$$

for $w \in W$, and $c \in N$, since $\overline{wc} = \overline{w} = w$.

The generators a_i are contained in N by definition, and thus we have

$$\sigma(wa_i) = wa_iw^{-1} \qquad (i = 1,\ldots,n).$$

For $x \in W$, put $A_x = xAx^{-1}$ and for $X \subseteq W$, put

$$A_X = \bigcup_{x \in X} A_x = \bigcup_{x \in X} xAx^{-1}.$$

Thus in particular

$$A_W = \{\sigma(wa_i) \mid i = 1, \ldots, n, \quad w \in W\}.$$

Letting S denote

$$S = \{\sigma(ws), \sigma(wt) \mid w \in W\} \sim \{1\},$$

we now have (in this notation) that

$$N = < S, A_W \mid - >,$$

the free group based on $S \cup A_W$. By definition

$$N_G = N/U = N/ < p,q >_F,$$

and

$$zpz^{-1} = \sigma(z)\overline{z}\, p\, \overline{z}^{-1}\sigma(z)^{-1}$$

for $z \in F$ (and similarly for zqz^{-1}). Thus

$$U = < p,q >_F$$

$$= < zpz^{-1}, zqz^{-1} \mid z \in F >$$

$$= < wpw^{-1}, wqw^{-1} \mid w \in W >_N,$$

and we have the following presentation of N_G:

$$N_G = < S, A_W \mid wpw^{-1}, wqw^{-1} \ (w \in W) >.$$

Consider next

$$M = < S >_N$$

and

$$M_G = < S_G >_{N_G}.$$

Observe that we may identify the quotient $T = N/M$ with $< A_W \mid - >$, and we have the commutative diagram

$$\begin{array}{ccc} N & \longrightarrow & N/M = T \\ \downarrow & & \downarrow \\ N_G & \longrightarrow & N_G/M_G = T_G. \end{array}$$

Denoting the images of wpw^{-1} and wqw^{-1} by, respectively,

$$p_w = wpw^{-1} \bmod M$$

$$q_w = wqw^{-1} \bmod M,$$

we obtain the following presentation for T_G:

$$T_G = N_G/M_G = < A_W \mid p_w, q_w \quad (w \in W) >. \qquad (3.4)$$

It is our aim to map the group T_G non-trivially to a group whose structure is known (e.g. a fundamental group of a graph of groups) and then deduce the freeness of $< A_G >$. In order to understand the presentation of T_G given in (3.4) we must analyze the defining relations in detail. We begin with the following lemma which easily follows by induction.

Lemma 3.1. *For all* $w \in W$ *and* $x \in < s,t \mid - > \sim \{1\}$ *we have that*

$$\sigma(wx) \in < S >.$$

We now prove a *translation of structure* lemma, which is key for our construction. Recalling the notation of definition 1, we compute the defining relators p_w and q_w explicitly by applying Lemma 3.1 and equations (3.1).

Lemma 3.2. *For* $w \in W$ *we have that*

$$wpw^{-1} \in < A_{\overline{wP}}, S \mid - >$$

$$wqw^{-1} \in < A_{\overline{wQ}}, S \mid - >.$$

Therefore, in $T = N/< S >_N = < A_W \mid - >$ *we have*

$$p_w \in < A_{\overline{wP}} \mid - > \quad \text{and} \quad q_w \in < A_{\overline{wQ}} \mid - >.$$

Proof. Recall that

$$p = x_1c_1 \ldots x_kc_k$$

where $\overline{x_i} \in < s,t >$, $\bar{x}_i \neq 1$, and $c_i \in < A > \sim \{1\}$, for $i = 1,\ldots,k$. Moreover

$p \in N = \langle A,p,q \rangle_F$.

Thus for $w \in W$, $wpw^{-1} \in N$,

$$wpw^{-1} = \sigma(wpw^{-1})$$

$$= \sigma(wx_1c_1 \dots x_kc_kw^{-1}).$$

Applying (3.1 (i)) we obtain

$$wpw^{-1} = \sigma(w)\sigma(\overline{w}\,x_1)\sigma(\overline{wx_1}\,c_1) \dots \sigma(\overline{wx_1 \dots c_{i-1}}\,x_i)\,\sigma(\overline{wx_1c_1 \dots x_i}\,c_i)$$

$$\dots \sigma(\overline{wx_1c_1 \dots x_kc_k}\,w^{-1}). \qquad (3.5)$$

Claim: $wpw^{-1} \in \langle A_{\overline{wP}}, S \mid - \rangle$.

For each $i = 1,\dots,k$ we have $c_i \in N$, and so

$$\overline{wx_1c_1 \dots x_i} = \overline{wx_1c_1 \dots x_ic_i} = \overline{wx_{(i)}}$$

where $x_{(i)} = x_1 \dots x_i$. Thus upon applying (3.3) we obtain

$$\sigma(\overline{wx_1c_1 \dots x_i}\,c_i) = (\overline{wx_{(i)}})c_i(\overline{wx_{(i)}})^{-1} \in \langle A_{\overline{wx_{(i)}}} \mid - \rangle.$$

Recalling Definition 2.1 we have that $P = \{x_{(1)}, \dots, x_{(k)}\}$ and thus

$$\overline{wx_{(i)}} \in \overline{wP}\,.$$

We conclude that $\sigma(\overline{wx_1c_1 \dots x_i}\,c_i) \in \langle A_{\overline{wP}} \mid - \rangle$ where again

$$A_{\overline{wP}} = \bigcup_{x(i) \in P} (\overline{wx_{(i)}})A(\overline{wx_{(i)}})^{-1}.$$

By applying Lemma 3.1 we obtain the inclusion

$$\sigma(\overline{wx_1c_1 \dots c_{i-1}}\,x_i) = \sigma(\overline{wx_{(i)}}) \in \langle S \rangle,$$

and similarly

$$\sigma(w) = 1, \;\; \sigma(\overline{wx_1c_1 \dots x_kc_k}\,w^{-1}) = \sigma(\overline{wx_{(k)}}\,w^{-1}) \in \langle S \rangle.$$

Combining the above data we deduce our claim.

It is clear that a similar computation may be executed for wqw^{-1}. Since by definition we have $p_w = wpw^{-1} \bmod \langle S \rangle_N$, it follows from (3.5) and the above calculations that

$$p_w = \sigma(\overline{wx_{(1)}}\, c_1) \ldots \sigma(\overline{wx_{(i)}}\, c_i) \ldots \sigma(\overline{wx_{(k)}}\, c_k) \in \langle A_{\overline{wP}} \mid - \rangle. \qquad (3.6)_p$$

Similarly we have

$$q_w = \sigma(\overline{wy_{(1)}}\, c_1) \ldots \sigma(\overline{wy_{(i)}}\, c_i) \ldots \sigma(\overline{wy_{(k)}}\, c_k) \in \langle A_{\overline{wQ}} \mid - \rangle. \qquad (3.6)_q$$

(Recall that here we are using the identification

$$N/\langle S \rangle_N = \langle A_W, S \mid - \rangle / \langle S \rangle_N = \langle A_W \mid - \rangle.)$$

It is the exact nature of the expressions $(3.6)_p$ and $(3.6)_q$ that allows us to now prove:

Lemma 3.3. *Assume that the following conditions hold:*

(a) $P \not\subset Q, Q \not\subset P$;

(b) $\overline{x_i} \neq 1$, $\overline{y_j} \neq 1$ *for* $i = 1,\ldots,k$ *and* $j = 1,\ldots,h$.

Then we have that:

(i) *The expressions* $(3.6)_p$ *and* $(3.6)_q$ *for* p_w, q_w *are reduced and cyclically reduced words relative to the generators* $A_{\overline{wP}}$, *and* $A_{\overline{wQ}}$, *respectively.*

(ii) *For* $w \in W$, *the natural homomorphisms*

$$\begin{array}{ccc} & & \langle A_{\overline{wP}} \mid p_w \rangle \\ & \nearrow i_p & \\ \langle A_{\overline{wP} \cap \overline{wQ}} \mid - \rangle & & \\ & \searrow i_q & \\ & & \langle A_{\overline{wQ}} \mid q_w \rangle \end{array}$$

are monic, and hence

$$\langle A_{\overline{wP}}, A_{\overline{wQ}} \mid p_w, q_w \rangle = \langle A_{\overline{wP}} \mid p_w \rangle *_{\langle A_{\overline{wP} \cap \overline{wQ}} \mid - \rangle} \langle A_{\overline{wQ}} \mid q_w \rangle. \qquad (3.7)$$

Consequently the canonical homomorphism induced by inclusion of generators

$$\langle A_1 \mid - \rangle \to \langle A_P, A_Q \mid p_1, q_1 \rangle$$

is monic.

Recall that R denotes the union $P \cup Q$.

Definition 3.4. For $w \in W$ define the group

$$\mathcal{G}(w) = \langle A_{\overline{wR}} \mid p_w, q_w \rangle$$

and for $w, w' \in W$ distinct, define

$$\mathcal{G}(w,w') = \langle A_{\overline{wR} \cap \overline{w'R}} \mid - \rangle.$$

Let

$$\alpha(w,w') : \mathcal{G}(w,w') \to \mathcal{G}(w)$$

denote the homomorphism induced by inclusion of generators.

In order for Definition 3.4 to define a graph of groups (on the complete graph with vertex set W), we require the following lemma:

Lemma 3.5. *Assume the defining relators* p,q *are independent relative to the generators* $\{s,t\}$. *Then the homomorphism*

$$\alpha_{(w,w')} : \mathcal{G}(w,w') \to \mathcal{G}(w) \text{ is monic for } w \neq w' \text{ in } W. \tag{3.8}$$

Lemma 3.5 follows from the generalized free product decomposition of the group $\mathcal{G}(w)$ given in Lemma 3.3, a technical analysis of generators of the group $\mathcal{G}(w,w')$, and the application of Magnus' Freiheitssatz to the groups $\langle A_{\overline{wP}} \mid p_w \rangle$ and $\langle A_{\overline{wQ}} \mid q_w \rangle$.

We may now prove our theorem.

Proof of Theorem 2.2. We are now in a position to consider the graph of groups $(\mathcal{G},\Gamma)$ where Γ denotes the complete graph based on the set W (i.e. the set of vertices of Γ is W, the set of edges of Γ is $W \times W$ and the terminal vertex of the edge (w_1,w_2) is just w_2), and the vertex and edge groups are given in Definition 3.4. In order to make some use of $(\mathcal{G},\Gamma)$ we begin by choosing a maximal subtree which suits our needs. Consider the subgraph of Γ spanned by the edges

$$\Delta = \{(\overline{wr}, w) \mid w \in W, r \in R \sim \{1\}\}. \tag{3.9}$$

Identifying W with F/N observe that by definition Δ is the generalized Cayley graph of

$$\langle R \sim \{1\} \rangle$$

acting by right multiplication by its image in F/N. By choosing a minimal (left) Schreier coset representative system $W_1 \subseteq F/N$ it can be shown Δ is the disjoint union, indexed by W_1, of copies of the ordinary Cayley graph of the group $< R \sim \{1\} >$ relative to its free basis $R \sim \{1\}$. This proves Δ is in fact a forest. Let Λ be the union

$$\Lambda = \bigcup_{z \in W_1} (1,z) \cup \Delta.$$

It is clear that Λ is a (rooted) tree which contains all the vertices of Γ, and thus Λ is in fact a maximal tree of Γ (for complete details see [A]). We record our observations in a lemma.

Lemma 3.6. *The graph Δ of (3.9) is a forest and is contained in a maximal subtree of Γ (denoted Λ).*

With the maximal subtree Λ now fixed we next consider the group

$$\pi_1(\mathfrak{G},\Gamma,\Lambda).$$

Our first move is to define a homomorphism

$$T = < A_W \mid - > \to \pi_1(\mathfrak{G},\Gamma,\Lambda)$$

which induces a non-trivial homomorphism

$$T_G = < A_W \mid p_w, q_w \ (w \in W) > \to \pi_1(\mathfrak{G},\Gamma,\Lambda).$$

Recall from the Bass-Serre theory (see [S]) that the natural homomorphisms

$$\phi_w : \mathfrak{G}(w) \to \pi_1(\mathfrak{G},\Gamma,\Lambda) \qquad (3.10)$$

are injective for all $w \in W$. Since $1 \in R$ we have that $w \in \overline{wR}$, and we may thus consider the composite set map

$$\{A_w\} \to \mathfrak{G}(w) \to \pi_1(\mathfrak{G},\Gamma,\Lambda). \qquad (3.11)$$

These functions induce a well defined mapping on the disjoint union

$$\phi : \bigcup_{w \in W} A_w \to \pi_1(\mathfrak{G},\Gamma,\Lambda).$$

Now suppose we have that $w \in \overline{vR}$ with $w \neq v$. In this situation we may also consider the composite set map

$$\{A_w\} \to \mathfrak{G}(v) \to \pi_1(\mathfrak{G},\Gamma,\Lambda). \qquad (3.12)$$

We claim that the functions (3.11) and (3.12) are identical. For suppose $w \in \overline{vR}$, and thus $w \in \overline{wR} \cap \overline{vR}$. This implies

$A_w \subseteq A_{\overline{wR}\cap\overline{vR}} \subseteq \mathfrak{G}(w,v).$

Recalling the fundamental properties of the fundamental group $\pi_1(\mathfrak{G},\Gamma,\Lambda)$ (see [S] for the details) we obtain the commutative diagram

$$\begin{array}{cccc} & & \mathfrak{G}(w) \;\rightarrow\; \pi_1(\mathfrak{G},\Gamma,\Lambda) & \\ & \nearrow & & \\ \mathfrak{G}(w,v) = \langle A_{\overline{wR}\cap\overline{vR}} \mid - \rangle & & \uparrow & \text{"conjugate by (w,v)"} \\ & \searrow & & \\ & & \mathfrak{G}(v) \;\rightarrow\; \pi_1(\mathfrak{G},\Gamma,\Lambda). & \end{array}$$

But notice that since $w \in \overline{vR}$, there is some $r \in R$, $r \neq 1$ such that $\overline{vr} = w$ and thus

$(w,v) = (\overline{vr}, v).$

Thus by our cunning choice of maximal trees

$(w,v) \in \Delta \subseteq \Lambda,$

and in the above diagram the mapping "conjugation by (w,v)" becomes trivial (again see [A, Section 6]). This leaves us with the commutative diagram

$$\begin{array}{ccccc} & & \mathfrak{G}(w) & & \\ & \nearrow & & \searrow & \\ \mathfrak{G}(w,v) & & & & \pi_1(\mathfrak{G},\Gamma,\Lambda) \\ & \searrow & & \nearrow & \\ & & \mathfrak{G}(v) & & \end{array}$$

Thus the restriction $\phi \downarrow A_w$ is also given by

$\{A_w\} \rightarrow \mathfrak{G}(v) \rightarrow \pi_1(\mathfrak{G},\Gamma,\Lambda)$

which was to be proved. Denoting the unique extension of ϕ to the free group again by ϕ

$\phi : T = \langle A_W \mid - \rangle \rightarrow \pi_1(\mathfrak{G},\Gamma,\Lambda)$

we now claim that ϕ induces a homomorphism on the quotient group

$T_G = N_G/M_G = \langle A_W \mid p_w, q_w \; (w \in W) \rangle,$

i.e. that $\phi(p_w) = \phi(q_w) = 1$ for all $w \in W$. This follows from the fact that $\phi \downarrow A_{\overline{wR}}$ is the composite of the homomorphisms

$$\{A_{\overline{wR}}\} \to \mathcal{G}(w) \to \pi_1(\mathcal{G},\Gamma,\Lambda)$$

and $\mathcal{G}(w) = \langle A_{\overline{wR}} \mid p_w, q_w \rangle$. Letting ϕ_G denote the homomorphism on T_G induced by ϕ we have

$$\begin{array}{ccccc} N & \to & T & & \\ \downarrow & & \downarrow & \searrow \phi & \\ N_G & \to & T_G & \xrightarrow{\phi_G} & \pi_1(\mathcal{G},\Gamma,\Lambda). \end{array}$$

Consider now the natural homomorphism from $\mathcal{G}(w)$ to T_G

$$\theta_w : \mathcal{G}(w) = \langle A_{\overline{wR}} \mid p_w, q_w \rangle \to T_G$$

making the following diagram commute:

$$\begin{array}{ccc} T_G & & \\ \uparrow & \searrow \phi & \\ \theta_w & & \pi_1(\mathcal{G},\Gamma,\Lambda). \\ & \nearrow \phi_w & \\ \mathcal{G}(w) & & \end{array}$$

Since ϕ_w is monic (see (3.9)), we have that θ_w is monic. Now recall from Lemma 3 that

$$\langle A_1 \mid - \rangle \to \mathcal{G}(1)$$

is monic. This implies the composite

$$\langle A_1 \mid - \rangle \to \mathcal{G}(1) \to T_G$$

is monic. We complete our argument by considering the commutative diagram

$$\begin{array}{ccc} \langle A_1 \mid - \rangle & & \\ \downarrow & \searrow & \\ N & & \mathfrak{G}(1) \\ \downarrow & & \downarrow \theta_1 \\ N_G & \rightarrow & T_G \end{array}$$

We deduce that the composite

$$\langle A_1 \mid - \rangle \hookrightarrow N \rightarrow N_G$$

is monic. But then so is the homomorphism

$$\langle A \mid - \rangle = \langle A_1 \mid - \rangle \rightarrow N_G \hookrightarrow G$$

of the proposition, which completes the proof.

Remark. The construction above actually produces a homomorphism

$$\phi_G : T_G \rightarrow \pi_1(\mathfrak{G},\Lambda),$$

to the fundamental group of the *tree* of groups. It can in fact be proved that ϕ_G is an isomorphism (see [A]), and this remark in conjunction with some technical analysis of the vertex groups and some general results on fundamental groups of graphs of groups allows one to deduce Theorem 2.3.

References

[A] Iris L Anshel, *A Freiheitssatz for a class of two relator groups* (Ph.D. Thesis, Columbia University 1989).

[B] G Baumslag, Groups with a single defining relation, in *Proceedings of Groups - St Andrews 1985* (ed. E F Robertson and C M Campbell, Cambridge University Press 1986).

[H] J Howie, How to generalize one relator group theory, in *Combinatorial Group Theory and Topology* (ed. S M Gersten and John R Stallings, Ann. of Math. Stud. **111**, Princeton University Press, 1987).

[L-S] Roger C Lyndon & Paul E Schupp, *Combinatorial Group Theory* (Springer-Verlag, Berlin, Heidelberg, New York, 1970).

[M-K-S] W Magnus, A Karrass & D Solitar, *Combinatorial Group Theory* (Second Revised Edition) (Dover Publications Inc., New York, 1976).

[S] J-P Serre, *Trees* (Springer-Verlag, Berlin, Heidelberg, New York, 1980).

AN INFINITE FAMILY OF NONABELIAN SIMPLE TABLE ALGEBRAS NOT INDUCED BY FINITE NONABELIAN SIMPLE GROUPS

ZVI ARAD

Bar-Ilan University, Ramat-Gan 92100, Israel

HARVEY BLAU

Northern Illinois University, DeKalb, Illinois 60115, USA

Recently, the present authors and E Fisman developed in [AB], [ABF] and [AF] a new theory of table algebras. This concept is defined as follows:

Definition 1. [AB] Let $\mathbf{B} = \{b_1 = 1_A, b_2, \dots, b_k\}$ be a basis of a finite dimensional, associative and commutative algebra A over the complex field C with identity element 1_A. Then $(A, \mathbf{B})$ is a *table algebra* (and $\mathbf{B}$ is a table basis) if the following hold:

(i) For all i,j,m, $b_i b_j = \sum_{m=1}^{k} b_{ijm} b_m$ with b_{ijm} a non-negative real number.

(ii) There is an algebra automorphism (denoted by $\bar{\ }$) of A whose order divides 2, such that $b_i \in \mathbf{B}$ implies that $\bar{b}_i \in \mathbf{B}$. (Then $\bar{i}$ is defined by $b_{\bar{i}} = \bar{b}_i$ and $b_i \in \mathbf{B}$ is called real if $b_i = \bar{b}_i$).

(iii) There is a function $g : \mathbf{B} \times \mathbf{B} \to \mathbf{R}^+$ (the positive reals) such that: $b_{ijm} = g(b_i, b_m) b_{\bar{j}mi}$, where b_{ijm}, $b_{\bar{j}mi}$ are defined in (i) and (ii) for all i,j,m.

Z Arad and E Fisman proved that for $(A, \mathbf{B})$ a table algebra there exists an algebra homomorphism $f : A \to \mathbf{C}$ such that $f(b_i) = f(\bar{b}_i) > 0$ for all $b_i \in \mathbf{B}$. (See [AB, Lemma 2.9].) Consequently, table algebras are equivalent to C-algebras with non-negative structure constants. (For the definition and details about C-algebras see [K], [H] and [BI].) Therefore theorems on table algebras yield new results for such C-algebras.

Examples of table algebras induced by finite groups

(1) Let G be a finite group. Let Ch(G) denote the set of all complex valued class functions on G, a commutative algebra under pointwise addition and multiplication. Let Irr(G) be the set of irreducible characters of G, a basis for Ch(G). Then (Ch(G),Irr(G)) satisfies (i)-(iii), where $^{-}$ extends linearly from $X_i \to \bar{X}_i$ (complex conjugate character) and $f(X_i) = X_i(1)$ for all $X_i \in$ Irr(G). Here, each λ_{ijm} is a non-negative integer and each $g(X_i, X_j) = 1$.

(2) Again, let G be a finite group and let Z(C(G)) be the centre of the complex group algebra. If $C_1 = \{1\}, C_2,\ldots,C_k$ are the conjugacy classes of G, let

$$b_i = \hat{C}_i = \sum_{g \in C_i} g$$

for each C_i, and let $Cla(G) = \{b_1,\ldots,b_k\}$. Then (Z(C(G)),Cla(G)) satisfies (i)-(iii), where $^{-}$ extends linearly from

$$b_i \to \bar{b}_i = \hat{C}_i^{-1} = \sum_{g \in C_i} g^{-1},$$

and $f(b_i) = |C_i|$. Here also each λ_{ijm} is a non-negative integer, but $g(b_i,b_m) = |C_i|/|C_m|$.

Examples of table algebras not induced by groups can be found in [AB] and [AF].

Consequently, theorems on table algebras can be applied to state theorems in finite group theory. We present in [AB], [ABF] and [AF] an abstract setting for the study of the decompositions of products of either irreducible characters or conjugacy classes of a finite group (treated together in a uniform way) and generalize results of Arad, Fisman, Chillag, Herzog, Lipman, Stavi and Mann which appeared in [AH], [AF1], [AF2], [AF3], [AF4], [AL] and [M].

A survey of the main results on table algebras can be found in [A].

Let $(A,\mathbf{B})$ be a table algebra of dimension k. If $a \in A$ and $b_i \in \mathbf{B}$, let $k(b_i,a)$ be the coefficient of b_i in a (for a is written as a linear combination of the elements of $\mathbf{B}$).

Definition 2. Let $(A,\mathbf{B})$ be a table algebra. The elements of $\mathbf{B}$ are called irreducible. If $a \in A$,

$$\mathrm{Irr}(a) := \{b_i \in \mathbf{B} \mid k(b_i,a) \neq 0\},$$

that is, Irr(a) is the support of a from $\mathfrak{B}$. The elements of Irr(a) are called the *irreducible constituents* of a.

Definition 3. A subset $\mathfrak{D} \subseteq \mathfrak{B}$ is called a *table subset* of $\mathfrak{B}$ if $\mathfrak{D} \neq \phi$ and $\mathrm{Irr}(b_i b_j) \subseteq \mathfrak{D}$ for all $b_i, b_j \in \mathfrak{D}$. Finally, $(A,\mathfrak{B})$ is *simple* if the only table subsets of $\mathfrak{B}$ are $\mathfrak{B}$ and $\{1\}$.

Definition 4. Let $(A,\mathfrak{B})$ be a table algebra. An irreducible element b_i is called *linear* if $\mathrm{Irr}(b_i^n) = \{1\}$ for some $n > 0$. $(A,\mathfrak{B})$ is called *abelian* if b_i is linear for all $b_i \in \mathfrak{B}$.

It is easy to see that if $\mathfrak{B} = \mathrm{Irr}(G)$, then $X \in \mathfrak{B}$ is linear in the above sense if and only if $X(1) = 1$; and if $\mathfrak{B} = \mathrm{Cla}(G)$ then $\hat{C}_i \in \mathfrak{B}$ is linear if and only if $|C_i| = 1$, i.e., $C_i \subseteq Z(G)$.

It is easy to verify that table algebras induced by nonabelian simple finite groups are nonabelian simple table algebras.

The next result may be viewed as a generalization of [FT, Theorem 2.6] and shows that table algebras not induced by groups are nonabelian.

Theorem A ([AB]). *If a table algebra* $(A,\mathfrak{B})$ *is abelian then there exists a finite abelian group* G *such that* A *is the group algebra* C[G} *and the elements of* $\mathfrak{B}$ *are positive real scalar multiples of the elements of* G.

Definition 5. The covering number $\mathrm{cn}(A,\mathfrak{B})$ of a table algebra, denoted simply by $\mathrm{cn}(\mathfrak{B})$, is the least positive integer m such that $\mathrm{Irr}(b_i^m) = \mathfrak{B}$ for all $b_i \neq 1$ in $\mathfrak{B}$, if such m exists.

Theorem B ([AB]). *Let* $(A,\mathfrak{B})$ *be a table algebra with* $|B| = k > 1$. *Let* r *be the number of real* $b_i \neq 1$ *in* $\mathfrak{B}$. *Then* $\mathrm{cn}(\mathfrak{B})$ *exists if and only if* $(A,\mathfrak{B})$ *is simple and nonabelian. If* $\mathrm{cn}(\mathfrak{B})$ *exists then*

$$\mathrm{cn}(\mathfrak{B}) \leq (k^2 - (r-1)^2)/2.$$

This result is Corollary 7.7 in Chapter 1 of [AH] in the special case $\mathfrak{B} = \mathrm{Irr}(G)$. It seems possible that there is an upper bound for $\mathrm{cn}(\mathfrak{B})$ which is linear in k, at least when $\mathfrak{B} = \mathrm{Irr}(G)$ or Cla(G). In fact for the sporadic groups, A_n, PSL(2,q), Sz(q) and groups of order less than 1,000,000, Arad, Dvir, Karni, Lipman and Ziser ([Ah], [AL], [Z]) found that $\mathrm{cn}(\mathrm{Irr}(G)) \leq k$ (not calculated for A_n) and also $\mathrm{cn}(\mathrm{Cla}(G)) \leq k$.

Our conjecture is that for a nonabelian simple group G, $\mathrm{cn}(\mathrm{Irr}(G)) \leq k$ and $\mathrm{cn}(\mathrm{Cla}(G)) \leq k$.

This conjecture is not valid in general for nonabelian simple table algebras and the bound of Theorem B is the best possible (at least for the real case) as the following example illustrates.

Example of an infinite family of nonabelian simple table algebras (A(k), 𝔅(k)) not induced by groups of dimension k, for k = 2,3,4,... which are real and with cn(𝔅) = 2k-2

We work in C[x], the polynomial ring over C.

Definition. $p_0 = 1$, $p_1 = x$. For $i \geq 1$,

$$p_{i+1} = p_1 p_i - p_{i-1}$$

(a Lucas sequence of polynomials).

Thus,

$$p_2 = p_1^2 - p_0 = x^2 - 1,$$

$$p_3 = p_1 p_2 - p_1 = x^3 - 2x,$$

$$p_4 = p_1 p_3 - p_2 = x^4 - 3x + 1, \text{ etc.}$$

Since $\deg p_i = i$, $\{p_i \mid i \geq 0\}$ is a basis for C[x]. Also, by definition, for all $j \geq 1$ we have

$$p_1 p_j = p_{j-1} + p_{j+1}. \qquad (*)$$

Proposition 1. *For all* $0 \leq i \leq j$,

$$p_i p_j = p_{j-i} + p_{j-i+2} + p_{j-i+4} + \dots + p_{j+i}.$$

Proof. This is clear for $i = 0$ and 1. For $i > 1$, we proceed by induction on i:

$$p_i p_j = p_j(p_1 p_{i-1} - p_{i-2}) = p_1(p_{i-1} p_j) - p_{i-2} p_j$$

$$= p_1(p_{j-i+1} + p_{j-i+3} + \dots + p_{j+i-1}) - (p_{j-i+2} + p_{j-i+4} + .. + p_{j+i-2}) \text{ (by induction)}$$

$$= p_{j-i} + p_{j-i+2} + p_{j-i+2} + p_{j-i+4} + \dots + p_{j+i-2} + p_{j+i} - (p_{j-i+2} + p_{j-i+4} + \dots + p_{j+i-2})$$

$$= p_{j-i} + p_{j-i+2} + p_{j-i+4} + \dots + p_{j+i} \quad \text{(by (*))}.$$

Definition. For all $a \in C[x]$, Irr(a) means the support of a from the basis $\{p_i \mid i \geq 0\}$.

Definition. For j any non-negative integer, let $\delta(j) = 0$ or 1, such that

$$\delta(j) \equiv j \pmod 2.$$

Proposition 2. *For all* $j \geq 0$,

$$\mathrm{Irr}(p_1^j) = \{p_j, p_{j-2}, p_{j-4}, \ldots, p_{\delta(j)}\}.$$

Proof. This follows by induction on j and (*).

Now fix $k \geq 2$. Define

$$q = q_k := p_1 p_k - p_{k-1} - p_k,$$

a polynomial of degree k+1. Then $p_{k+1} = p_1 p_k - p_{k-1} \equiv p_k \pmod q$, and by induction on j (since $p_{k+j+1} = p_1 p_{k+j} - p_{k+j-1}$),

$$p_{k+j+1} \equiv p_{k-j} \pmod q, \quad \text{for } 0 \leq j \leq k. \qquad (**)$$

From Proposition 2 and (**) we have

Proposition 3. *For* $k < j \leq 2k$,

$$\mathrm{Irr}(p_1^j) \equiv \{p_{2k-j+1}, p_{2k-j+3}, \ldots, p_{k-\delta(j+k)-1}, p_k, p_{k+\delta(j+k)-2}, p_{k+\delta(j+k)-4}, \ldots, p_{\delta(j)}\}$$
(mod q).

In particular,

$$\mathrm{Irr}(p_1^{2k-1}) \equiv \{p_2, p_4, p_6, \ldots, p_{k-2+\delta(k)}, p_k, p_{k-\delta(k)-1}, p_{k-\delta(k)-3}, \ldots, p_3, p_1\}$$

and

$$\mathrm{Irr}(p_1^{2k}) \equiv \{p_1, p_3, \ldots, p_{k-\delta(k)-1}, p_k, p_{k+\delta(k)-2}, p_{k+\delta(k)-4}, \ldots, p_2, p_0\} \pmod q.$$

Now let $A = \mathbb{C}[x]/(q)$, b_i = image of p_i in A, $0 \leq i \leq k$. Let $\mathbf{B} = \{b_0, b_1, \ldots, b_k\}$. Then dim A = k+1, and $\mathbf{B}$ is a basis for A. It follows from Proposition 1 and (**) that $(A,\mathbf{B})$ satisfies properties (i), (ii) and (iii) of [AB], where $^-$ is defined as the identity map. Hence $(A,\mathbf{B})$ is a real table algebra. Furthermore, all the multiplication constants are either 1 or 0, and $(A,\mathbf{B})$ is normalized, in the sense of [AB].

Again by Proposition 1, for any i with $0 < i \leq k$, p_k or p_{k+1} is in $\mathrm{Irr}(p_i^n)$ for sufficiently large n. Thus $b_k \in \mathrm{Irr}(b_i^n)$. Now $\mathrm{Irr}(b_k^2) = \mathbf{B}$, by Proposition 1 and (**). It follows that $(A,\mathbf{B})$ is simple.

From Proposition 3, $\mathrm{Irr}(b_1^{2k-1}) \subsetneq \mathbf{B} = \mathrm{Irr}(b_1^{2k})$. So by [AB, Theorem B], $\mathrm{cn}(\mathbf{B}) = 2k$.

Therefore, $(A,\mathbf{B})$ is a real, simple, nonabelian table algebra with $\mathrm{cn}(\mathbf{B}) = 2(|\mathbf{B}|-1)$.

Proposition 4. *$(A,\mathbf{B})$ as above cannot be obtained as a rescaling, in the sense of* [AB], *of* (Ch(G),Irr(G)) *or* (Z(CG),Cla(G)) *for any finite group* G.

Proof. Suppose that $\mathbf{B} = \{b_0,b_1,\ldots,b_k\}$ as above occurs as a rescaling of Irr(G) for some (necessarily simple, nonabelian) G. Since $\mathbf{B}$ is normalized, we have $\mathbf{B} = \mathrm{Irr}(G)$ and $b_1 = X$ for some $X \in \mathrm{Irr}(G)$. Now $b_1^2 = b_0+b_2$ implies that $b_1^2-b_0 \in \mathrm{Irr}(G)$. But then the symmetric and skew decomposition of X^2 forces 1_G to be the skew component of X^2. So $X(1) = 2$. But the (ancient) classification of complex linear groups of degree two tells us that G cannot be nonabelian simple. This is a contradiction.

Suppose that $\mathbf{B}$ occurs as a rescaling of Cla(G) for some (simple, nonabelian) G. Then $b_i = (1/\sqrt{|C_i|})\hat{C}_i$ for conjugacy classes C_i, and $b_i^2 = b_0+b_2$ implies that

$$\hat{C}_1^2 = |C_1|1 + (|C_1|/\sqrt{|C_2|})\hat{C}_2. \qquad (***)$$

But then $\sqrt{|C_2|}$ must be an integer factor of $|C_1|$, and (***) yields also that

$$|C_1|^2 = |C_1|+|C_1|\cdot\sqrt{|C_2|}.$$

Hence $|C_1| = 1 + \sqrt{|C_2|}$. It follows that $|C_2| = 1$ and $|C_1| = 2$, which contradicts the simplicity of G.

An example of a theorem in table algebra theory

In [ABF] we stated the following general theorem about table algebras which presents the possibilities of a table algebra theory.

Theorem C. *Let $(A,\mathbf{B})$ be a nonabelian simple table algebra. Then*

$$\mathbf{B} = \mathrm{Irr}\Big(\prod_{b\in\mathbf{B}} b\Big) \cup \{1\}.$$

Theorem C implies the following two results as corollaries:

(a) *In a nonabelian finite simple group* G, *the product of all the distinct irreducible characters of* G *contains all the nontrivial irreducible characters of* G *as constituents.*

(b) *In nonabelian finite simple groups* G, *the product of all the distinct conjugacy classes of* G *contains* G-{1}.

The first result (a) is a solution to an open problem in finite group theory. It is still unknown if 1_G is always a constituent. Assuming the classification theorem for the finite nonabelian simple groups, this has been checked [ABF] for all cases except A_n, the infinite family of the alternating groups on n points.

The second result (b) is known as Brauer's theorem which states that the product of all the distinct conjugacy classes of a nonabelian simple group G is exactly G.

It is easy to constuct an example of a nonabelian simple table algebra $(A,\mathbf{B})$ of dimension 4, $\mathbf{B} = \{1,b,\bar{b},c\}$ s.t. $\mathrm{Irr}(b\bar{b}c) = \{b,\bar{b},c\} = \mathbf{B} - \{1\}$. Thus the result of Theorem C is best possible.

Let us conclude with a constructive method of producing new table algebras not induced by groups.

New examples of table algebras not induced by groups

Definition 6 [AF]. Let $(A,\mathbf{B})$ be a table algebra and $\mathbf{N} = \{a_i \in \mathbf{B} \mid i \in I\}$ a table subset of $\mathbf{B}$. Then we will say that $(\mathbf{B},\mathbf{N})$ is a generalized CF-pair if $a_i a_j = \lambda_{ijj} a_j$ for every $j \notin I$ and $i \in I$.

Definition 7 [AF]. Let $(A,\mathbf{B})$ and $(A',\mathbf{B}')$ be table algebras such that A and A' are isomorphic algebras, $\mathbf{B} = \{b_1,\ldots,b_k\}$ and $\mathbf{B}' = \{b'_1,\ldots,b'_k\}$, table bases of A and A', respectively, with $b_i b_j = \sum_{\ell=1}^{k} b_{ij\ell}\, b_\ell$ and $b'_i b'_j = \sum_{\ell=1}^{k} b'_{ij\ell} b'_\ell$. Then we will say that $\mathbf{B}$ and $\mathbf{B}'$ are similar bases if there exists a permutation π of the set $\{1,\ldots,k\}$ such that $b_{ij\ell} = b'_{\pi(i)\pi(j)\pi(\ell)}$ for every i,j,ℓ.

In [AF, Lemma 2.14] we proved the following:

Given any two table algebras $(B,\mathbf{B})$ *and* $(C,\mathbf{C})$ *of dimension* n *and* m, *respectively, there exists always a unique table algebra* $(A,\mathcal{A})$ *(up to permutation of the elements of* $\mathcal{A}$*) of dimension* $k = n+m-1$ *such that the following properties hold:*

(i) *Let* $\mathbf{N} = \{a_i \in \mathcal{A} \mid 1 \le i \le n\}$ *then* $\mathbf{N}$ *is a table subset of* $\mathcal{A}$.

(ii) $\mathbf{N}$ *and* $\mathbf{B}$ *are similar bases.*

(iii) $(\mathcal{A},\mathbf{N})$ *is a generalized* CF*-pair.*

(iv) *See* [AF, Lemma 2.14]. (We need more background and definitions in order to state this property.)

More detailed information on $(A,\mathcal{A})$ is available in [AF].

In particular we have a constructive way to produce a new table algebra (not necessarily induced by a group) from any two table algebras induced by groups. Finally, the examples we have of nonabelian simple table algebras induced and noninduced by groups give us the feeling that classifying all nonabelian simple table algebras is not a realistic dream. But there is still much to study about table algebras.

Acknowledgements

This research was supported by the United States-Israel Binational Foundation Grant No. 86-0049.

Part of this paper was completed while the first author was a visitor at the Universita' Degli Studi Di Trento. He wishes to express his gratitude for their hospitality.

References

[A] Z Arad, Survey on table algebras and applications to finite group theory, *Israel Mathematical Conference Proc.* (The Weizmann Science Press of Israel **1** 1989), 96-110.

[AB] Z Arad & H Blau, On table algebras and applications to finite group theory, *J. Algebra*, to appear.

[ABF] Z Arad, H Blau & E Fisman, On table algebras and applications to finite group theory, *J. Algebra*, to appear.

[AF] Z Arad & E Fisman, On table algebras, C-algebras and applications to finite group theory, submitted.

[AF1] Z Arad & E Fisman, An analogy between products of two conjugacy classes and products of two irreducible characters in finite groups, *Proc. Edinburgh Math. Soc.* **30** (1987), 7-22.

[AF2] Z Arad & E Fisman, About products of irreducible characters and products of conjugacy classes in finite groups, *J. Algebra* **114** (1988), 466-476.

[AF3] Z Arad & E Fisman, On products of irreducible characters and conjugacy classes of finite groups in complementary sets, *Proc. Sympos. Pure Math.* **47** (1987), 11-23.

[AH] Z Arad & M Herzog, eds. *Products of Conjugacy Classes in Groups* (Lecture Notes in Math. **1112**, Springer-Verlag, Berlin, 1985).

[AL] Z Arad & H Lipman-Gutweter, On products of characters in finite groups, *Houston J. Math.* **15** (1989), 305-326.

[BI] E Bannai & T Ito, *Algebraic Combinatorics I: Association Schemes* (The Benjamin/Cummings Pub. Co., Inc., Menlo Park, 1984).

[Br1] R Brauer, Some applications of the theory of blocks of characters of finite groups, II, *J. Algebra* **1** (1964), 307-334.

[Br2] R Brauer, On pseudo groups, *J. Math. Soc. Japan* **20** (1968), 13-22.

[Ca] A R Camina, Some conditions which almost characterize Frobenius groups, *Israel J. Math.* **31** (1978), 153-160.

[FT} P A Ferguson & A Turull, Algebraic decompositions of commutative association schemes, *J. Algebra* **96** (1985), 211-229.

[H] G Hoheisel, Uber Charaktere, *Monatsh. Math. Phys.* **48** (1939), 448-456.

[K] Y Kawada, Uber den Dualitatssatz der Charaktere nichtcommutativer Gruppen, *Proc. Phys. Math. Soc. Japan* (3) **24** (1942), 97-109.

[M] A Mann, Products of classes and characters in finite groups, *J. Algebra*, to appear.

[Z] I Zisser, The covering numbers of the sporadic simple groups, *Israel J. Math.*, to appear.

HORACE Y MOCHIZUKI : IN MEMORIAM

SEYMOUR BACHMUTH

University of California at Santa Barbara, California CA 93106, USA

Bibliographical notes:

Horace Mochizuki was born in California in 1937. He grew up on a farm in the Great Central Valley. He was educated at the University of Redlands and received his Ph.D. from the University of Washington, Seattle. He has been on the faculty of the University of California at Santa Barbara since 1965.

In 1985 I stood here and delivered a series of lectures on the work of Horace and myself. It is a sad occasion on which I address you again today, sharing some thoughts of Horace.

I mention also my sadness at the passing of Roger Lyndon the previous year. Two of the most important mathematical influences in my life are gone, but more important, the personal loss is overwhelming.

St Andrews is a fitting place to share memories of Horace, as he was closely involved with Groups-St Andrews. St Andrews had many pleasant associations for both of us. We attended the first conference in 1981, and it was immediately after that meeting that Horace made a crucial mathematical discovery which I will later discuss. He did not attend Groups-St Andrews 1985, but as already mentioned, I spoke extensively about our then recent work. We had both looked forward to this meeting until Horace became ill.

I met Horace in 1965, when we both came to the University of California in Santa Barbara after two years of postgraduate work. In December of that year we began our long collaboration. I hardly need mention to this audience how fortunate I am to have known Horace personally as well as professionally.

Horace was a ring theorist by training. His first three published papers have the titles: Finitistic global dimensions for rings; On the double commutator algebra of QF-3 algebras; A characterization of QF-3 rings. His fourth publication was our first collaboration. The nineteenth paper on his publication list, a 1978 publication in the Canadian Mathematical Bulletin was titled: "Unipotent matrix groups over division rings". In it he proved what one might call 'The non-commutative Kolchin Theorem'.

If G *is a unipotent group of* $n \times n$ *matrices over a division ring of characteristic* 0 *or prime* $p > (n-1)\,[n/2]$ *where* $[n/2]$ *is the greatest integer less than or equal to* $n/2$, *then* G *can be simultaneously triangularized.*

In 1977, friends, colleagues, and students of Ellis Kolchin joined together to present a volume of mathematical papers to him on the occasion of his sixtieth birthday. In this Academic Press publication, several of the papers contained applications and/or new proofs of Kolchin's theorem in various formulations (see e.g., the articles by Steinberg, Tits and Kaplansky). An immediate observation is that all the variations and generalizations involve fields. The article by Kaplansky has the title "The Engel-Kolchin Theorem Revisited". I quote from this article:

> Beyond the theorems discussed here, there lies an uncharted region where finiteness assumptions are dropped. Here is an example of the kind of statement that can be proposed (I do not say 'conjectured', feeling that to be too daring): in a ring with unit, let S be a multiplicative semigroup consisting of unipotent elements; then the nilpotent parts of the members of S generate a nil subring. A special case is the assertion that the Engel-Kolchin theorem works over a division ring.

Thus, Horace not only ventured into an uncharted region, but made a definitive contribution to the subject.

In many of Horace's papers one finds theorems or crucial computations in rings be they group rings, Lie rings, polynomial rings, matrix rings, rings of characteristic zero, rings of prime characteristic, rings of one sort or another. But apart from his thesis and first three papers, most, if not all, were motivated and established to prove results about groups. This was the case in his paper on unipotent matrix groups. Another good illustration was our joint paper with Walkup concerning groups of exponent 5. There the main theorem was about associative rings. But probably no one regarded that except for its application to groups. Despite his wide and very impressive knowledge of general algebra, Horace considered himself a group theorist. In his mind there was never any question about that.

Horace's final research effort was last fall while on sabbatical at MSRI in Berkeley. The previous summer he had participated in the group theory conference in Korea, the proceedings of which will contain our final collaboration. It is ironic that this last joint work drew so heavily on material from our beginning work. It also turns out that our first collaboration played an important symbolic role in his very last work done at MSRI. Let me briefly recall this beginning:

Consider a free metabelian group of rank $n > 2$ *and let* $G(k)$ *denote its nilpotent quotient of class* $k > 3$. *A matrix* I_n+A *(of a special form), with entries in the integral group ring of the free abelian group of rank* n *modulo the* k-2 *power of the augmentation ideal, represents an automorphism of the group* $G(k)$. *This lifts to an automorphism of the corresponding free metabelian group if and only if* A *has trace zero.*

I will later refer to this as the trace theorem. This result from my thesis turned out to be of exceptional importance to me because it converted Horace into a group theorist! Horace admired this theorem for reasons not easily explained here.

Generalizations involving matrices over more general group rings were beckoning. But, as you can well imagine, situations involving automorphism groups can become strange and chaotic when zero divisors are present. Horace became intrigued, and he understood better than I the possibilities as well as the pitfalls. This first collaboration generalized some of the results to certain groups with torsion. Our last collaboration, which I will discuss later, brought us back to this beginning.

After this first joint publication in automorphism groups, both of us turned our attention to questions centred around the Burnside problem. Looking at Horace's publication list, I count ten listings all told in this area, not including the recent reprinting of our joint paper with H A Heilbronn in his Collected Works and our 1970 AMS Bulletin announcement. Nine of these ten are collaborative efforts. I am involved in eight, Ken Weston in two, while Narain Gupta, Hans Heilbronn and David Walkup are each involved in one. I hardly need detail Horace's contributions to this subject. They are already expertly documented in the lecture notes of Sean Tobin in the proceedings of Groups-St Andrews 1981 as well as in the recent books on the Burnside problem by Kostrikin and Vaughan-Lee. There are many younger people in the audience who are only aware of Horace's more recent work in automorphism groups. In speaking here, however briefly, of his work, it would be unfair not to convey the impact and excitement generated by his Burnside research. To give you some inkling of this, George Glauberman wrote in the mid seventies that, at least in his recollection, the two papers ([10] and [11] on Horace's publication list) were the only papers ever studied in their Chicago seminar which were not in finite group theory.

After this Burnside excursion Horace returned to problems involving automorphism groups. Initially he and I together considered problems centred around the Tits alternative. We were dealing with matrices over division rings, and this motivated Horace to consider the problem of the non-commutative Kolchin Theorem. Our joint research during this period culminated in a paper we published in the Communications in Algebra with the title "Automorphism groups and subgroups of SL_2 over division rings".

The developments in that paper led to our 1978 Journal of Algebra article, in which we proved that the IA-automorphism group of the free metabelian group of rank 3 was infinitely generated. The story that developed beginning with that paper and a 1977 paper by Suslin was the subject of my 1985 lectures. These lectures were summarized in the two part expository papers written for the Groups-St Andrews 1985 Proceedings, the first part written by myself and the second by Horace. Horace played an indispensable role in the preparation of those lectures. Although not actually there, I can safely say that Horace's presence was felt at Groups-St Andrews 1985.

It was after the 1981 St Andrews meeting, while travelling with Gerhard Rosenberger on a train to London, that Horace discovered that the following automorphism (or a similar one) of the 3 generator metabelian group

$x \to x[y,z,x,x]$

$y \to y$

$z \to z$

lifts to a free automorphism in rank four. This turned out to be the breakthrough we were looking for. After proving that the 3 generator metabelian group is infinitely generated, we had thought the result automatic for all larger ranks. Of course the proof wasn't there. We were puzzled and only Suslin's recent work on SL_n for $n > 2$, together with our work on SL_2 gave us pause that something strange might be in the air. The breakdown of one simple automorphism, although a far cry from the full theorem, was enough to liberate our thinking and give us direction. One immediately wondered whether the tame automorphisms could conceivably be a normal subgroup, etc. Even though, at that time, we had no realistic expectation of being able to prove such results, the thought itself was revolutionary as the history of K-theory demonstrates.

The resulting theorem that all automorphisms of the free metabelian group of large enough rank are induced from free automorphisms raised the question: Is it ever the case that automorphism groups can be non-finitely generated in more than a finite number of ranks? Our final collaboration, to appear in the proceedings of the 1988 Korea conference, showed the existence of a free group in a solvable variety whose automorphism group is infinitely generated in every (non-trivial) finite rank. The proof made use of results contained in recent joint work that we had done with Gilbert Baumslag and Joan Dyer, but the crucial ideas that we needed came from our first collaboration.

To describe Horace's final work in Berkeley, let me remind you first of a theorem of Elena Stohr. Using our result that all automorphisms of the free metabelian group of ranks larger than 3 are tame (that is induced by automorphisms of the same rank free group), she was able to show:

The automorphism group of the free centre-by-metabelian group of ranks larger than 3 is generated by tame automorphisms together with the single automorphism

$a \to a[[a,b],[c,d]]$

$b \to b$

$c \to c$

$d \to d.$

Whether or not this automorphism is tame in ranks four or larger was unknown. One of the corollaries of Horace's work at MSRI was that this special automorphism is not even stably tame. This was also proved recently and jointly by Roger Bryant, Kanta Gupta and Frank Levin. The two proofs were different, but they are combining in a joint publication.

The final work of Horace at MSRI was motivated by the problem of when an automorphism of a free nilpotent group G(n,k) of rank n and class k is induced by an automorphism of the free group of rank n. Let K(n,k) denote the kernel of the natural map of the automorphism group of G(n,k) into the automorphism group of G(n,k-1), so that K(n,k) is a finitely generated free abelian group whose rank can be computed. The rank of the subgroup K*(n,k) of those automorphisms in K(n,k) which can be lifted to free automorphisms is another story. For example, using our theorem that all automorphisms of the free metabelian group of rank larger than 3 can be lifted, together with the trace theorem mentioned earlier, Horace and I were able to deduce that the rank of K*(n,4) is $2n\binom{n+1}{n-2} - \binom{n+1}{n-1}$ for all $n > 2$. Horace wanted to produce such a result for K*(n,k), $k > 4$. For starters, he was looking for a condition analogous to the trace theorem. The situation, however, is harder and much more complex. Horace found some necessary conditions, depending on when Fox derivatives of certain elements of a relevant group ring are pure Lie elements, and was working toward the establishment of sufficient conditions. He was scheduled to discuss this material in April in our seminar in Santa Barbara. I am sure that he would have spoken on this work in the address he was invited to give at Groups-St Andrews 1989. It was not to be.

It was at the very end of March, during the quarter break, that Horace learned of his illness. He was very brave when he fought for his life, showing more concern for others than for himself.

I end these remarks about Horace with the following poem.

I have said that the soul is not more than the body,

And I have said that the body is not more than the soul,

And nothing, not God, is greater to one that one's self is,

And whoever walks a furlong without sympathy walks to his own funeral drest in his shroud,

And I or you pocketless of a dime may purchase the pick of the earth,

And to glance with an eye or show a bean in its pod confounds the learning of all times,

And there is no trade or employment but the young man following it may become a hero,

And there is no object so soft but it makes a hub for the wheeled universe.

Walt Whitman

Song of Myself.

Publication List

Horace Y Mochizuki

1. Finitistic global dimensions for rings, *Pacific J. Math.* **15** (1965), 249-258.
2. On the double commutator algebra of QF-3 algebras, *Nagoya Math. J.* (1965), 221-230.
3. A characterization of QF-3 rings (with L E T Wu and J P Jans), *Nagoya Math. J.* (1966), 7-13.
4. Automorphisms of a class of metabelian groups II (with S Bachmuth), *Trans. Amer. Math. Soc.* **127** (1967), 294-301.
5. Cyclotomic ideals in groups rings (with S Bachmuth), *Bull. Amer. Math. Soc.* **72** (1966), 1018-1020.
6. Metabelian Burnside groups (with S Bachmuth and H Heilbronn), *Proc. Roy. Soc. London Ser. A.* **307** (1968), 235-250.

6(a). Metabelian Burnside groups (with S Bachmuth and H Heilbronn), reprinted in the *Collected Works of Hans Arnold Heilbronn* (edited by E J Kani and R A Smith, John Wiley and Sons).

7. Kostrikin's theorem on Engel groups of prime power exponent (with S Bachmuth), *Pacific J. Math.* **26** (1968), 197-213.
8. The class of the free metabelian group with exponent p^2 (with S Bachmuth), *Comm. Pure Appl. Math.* **21** (1968), 385-399.

9. A nonsolvable group of exponent 5 (with S Bachmuth and D Walkup), *Bull. Amer. Math. Soc.* **76** (1970), 638-640.

10. Construction of a nonsolvable group exponent 5 (with S Bachmuth and D Walkup), *Word Problems* (edited by W Boone, R Lyndon and F Canonito, North Holland Publishing Company, Amsterdam, 1973).

11. Third Engel groups and the Macdonald-Neumann conjecture (with S Bachmuth), *Bull. Austral. Math. Soc.* **5** (1971), 379-386.

12. On groups of exponent 4 with generators of order 2 (with N Gupta and K Weston), *J. Austral. Math. Soc.* **10** (1974), 135-142.

13. A group of exponent 4 with derived length at least 4 (with S Bachmuth and K Weston), *Proc. Amer. Math. Soc.* **39** (1973), 228-234.

14. A criterion for nonsolvability of exponent 4 groups (with S Bachmuth), *Comm. Pure Appl. Math.* **26** (1973), 601-608.

15. On groups of exponent 4: A criterion for nonsolvability, *Proc. Second Intl. Conference Theory of Groups* (Canberra, 1973), 499-503.

16. Automorphisms of Solvable Groups (with S Bachmuth), *Bull. Amer. Math. Soc.* ***81*** (1975), 420-422.

17. IA-automorphisms of two-generator torsion-free groups (with S Bachmuth and E Formanek), *J. Algebra* **40** (1976), 19-30.

18. Triples of 2×2 matrices which generate free groups (with S Bachmuth), *Proc. Amer. Math. Soc.* **59** (1976), 25-28.

19. Unipotent matrix groups over division rings, *Canad. Math. Bull.* **21** (2) (1978), 249-250.

20. IA-automorphisms of the free metabelian group of rank 3 (with S Bachmuth), *J. Algebra* **55** (1978), 106-115.

21. Automorphism groups and subgroups of SL_2 over division rings (with S Bachmuth), *Comm. Algebra* **7** (14) (1979), 1531-1558.

22. $E_2 \neq SL_2$ for most Laurent polynomial rings (with S Bachmuth), *Amer. J. Maths.* ***104*** (1982), 1181-1189.

23. The non-finite generation of Aut(G), G free metabelian of rank 3 (with S Bachmuth), *Trans. Amer. Math. Soc.* **270** (1982), 693-700.

24. GL_n and the automorphism groups of free metabelian groups and polynomial rings (with S Bachmuth), *Groups - St Andrews 1981* (edited by C M Campbell and E F Robertson, Lecture Note Series of London Math. Soc. **71**, Cambridge University Press, Cambridge, 1982), 160-168.

25. The finite generation of Aut(G), G free metabelian of rank ≥ 4 (with S Bachmuth), *Contemp. Math.* **33** (1984), 79-81.

26. Aut(F) → Aut(F/F") is surjective for free group F of rank ≥ 4 (with S Bachmuth), *Trans. Amer. Math. Soc.* **292** (1985), 81-101.

27. Lifting nilpotent automorphisms and the (3,5) theorem (with S Bachmuth), *Arch. Math.* **47** (1986), 103-106.

28. Automorphism groups of 2-generator metabelian groups (with S Bachmuth, G Baumslag and J Dyer), *J. London Math. Soc.* (2) **36** (1987), 393-406.

29. Automorphisms of solvable groups, Part II, *Groups-St Andrews 1985*, (edited by E F Robertson and C M Campbell, Lecture Note Series of London Math. Soc. **121**, Cambridge University Press, Cambridge, 1987), 15-29.

30. Can automorphism groups ever be infinitely generated in more than a finite number of dimensions? (with S Bachmuth), *Adv. in Math.* **76** (1989), 218-229.

31. The tame range of automorphism groups and GL_n (with S Bachmuth), *Proceedings of the 1987 Singapore Group Theory Conference*, (edited by K N Cheng and Y K Leong, Walter de Gruyter & Co. Berlin-New York, 1989), 241-251.

32. The infinite generation of automorphism groups (with S Bachmuth), *Groups Korea 1988*, (Springer Lecture Series, **1398**, 1989), 25-28.

33. Non-tame automorphisms of free nilpotent groups (with R M Bryant, C K Gupta and F Levin), preprint.

BOUNDS ON CHARACTER DEGREES AND CLASS NUMBERS OF FINITE NON-ABELIAN SIMPLE GROUPS

EDWARD A BERTRAM

University of Hawaii at Manoa, Honolulu, HI 96822, USA

MARCEL HERZOG

Tel-Aviv University, Tel-Aviv, Israel

Abstract

Using Jordan's theorem on the index of an abelian normal subgroup of maximum order in a linear group, together with a precise bound on $\pi(x)$, we prove a lower bound on the degrees of non-principal characters of finite non-abelian simple groups G. Corollaries include an upper bound for the number of conjugacy classes of G and substantial improvements in several inequalities relating |G| and the number of involutions in G.

Let G be a finite group of order |G| having k(G) conjugacy classes and distinct irreducible complex characters with degrees $\chi_1(1) = 1, \chi_2(1),...,\chi_k(1)$. Using deep results of R Brauer in the theory of modular characters, W Feit and J G Thompson [5] showed that if the prime p | |G| and a Sylow p-subgroup is not normal in G, then $\chi(1) \geq \frac{p-1}{2}$ for each faithful character χ of G. In particular, if G is a finite non-abelian simple group and χ is any non-principal character of G, then $\chi(1) \geq \frac{p-1}{2}$ for the largest prime p | |G|. In this paper we give a lower bound for $\chi(1)$ in terms of the order of G when G is simple and use this to give improved bounds on k(G) and the number of involutions in G.

Theorem. *If* G *is a finite non-abelian simple group and* χ *a non-principal character of* G, *then*

$$\chi^2(1) > \frac{9}{71} \ln |G| \ln \ln |G|.$$

Proof. Our proof depends on C Jordan's theorem in the form proved by Blichfeldt [3, p.177]: Let G be a finite subgroup of GL(n,C), n ≥ 2. Then there exists an abelian normal subgroup A ≤ G such that

$$[G : A] < n!(6^{(n-1)})^{\pi(n+1)+1}.$$

Here $\pi(n+1)$ is the number of primes $\leq n+1$, for which we use the precise estimate given in [6, p.69 (3.6)]:

$$\pi(n+1) < 1.25506 \frac{n+1}{\ln(n+1)} \quad (n \geq 1).$$

Let Irr(G) denote the set of all the irreducible complex characters of G. Since G is simple, each of the k(G) - 1 non-principal characters $\chi \in \mathrm{Irr}(G)$ is faithful and G is isomorphic to a subgroup of GL(n,C), where $n = \chi(1) \geq 3$. By Jordan's theorem we have $|G| < n!(6^{n-1})^{\pi(n+1)+1}$. We claim that

$$n!(6^{n-1})^{\pi(n+1)+1} < 16^{n^2/\ln n}, \text{ for } n \geq 3. \qquad (1)$$

First (1) is checked for $3 \leq n \leq 26$, and we assume $n \geq 27$. Next $n!6^{n-1} < n^{5n/4}$ for $n \geq 27$ is proved by induction, and finally

$$n^{5n/4} < \left(\frac{5}{3}\right)^{n^2/\ln n}$$

follows from the fact that $x/(\ln x)^2$ is an increasing function for $x > e^2$. Thus

$$n!6^{n-1} < \left(\frac{5}{3}\right)^{n^2/\ln n}$$

for $n \geq 27$. Using the above-mentioned upper bound for $\pi(n+1)$ we see that

$$(n-1)\pi(n+1) < 1.256n^2/\ln n$$

and thus

$$n!(6^{n-1})\cdot 6^{(n-1)\pi(n+1)} < 16^{n^2/\ln n}$$

for all $n \geq 27$. Now inequality (1) is true in all cases, and by Jordan's theorem $|G| < 16^{n^2/\ln n}$. From this it follows that

$$\ln|G| \ln\ln|G| < n^2 \left\{ 2 + \frac{\ln\ln 16 - \ln\ln n}{\ln n} \right\} \ln 16.$$

Since the function $\frac{\ln\ln 16 - \ln\ln x}{\ln x}$, defined for $x \geq 3$, is maximized when $x = 3$, we see that $\ln|G| \ln\ln|G| < \frac{71}{9} n^2 = \frac{71}{9}\chi^2(1)$ for each non-principal character $\chi \in \mathrm{Irr}(G)$.

Remarks. The factor ln ln |G| in the lower bound of the theorem cannot be replaced by ln |G|, i.e. $\chi(1) > c \ln|G|$ is not a theorem for any constant $c > 0$. For the group Alt(n) of all even permutations of $\{1,2,3,\ldots,n\}$, $n \geq 4$, is doubly transitive and hence [3, Theorem 11.3 (d)] the permutation character is the sum of the principal character and an irreducible character χ of degree n-1. Since

$$|Alt(n)| = \frac{n!}{2} > \left(\frac{n}{e}\right)^n,$$

it is easy to see that $\chi(1) = n-1 > c \ln |Alt(n)|$ is false for all sufficiently large n (depending on c).

For any finite group G with commutator subgroup G' we know that [G : G'] of the $\chi_i(1)$ are equal to 1,

$$\sum_{i=1}^{k} \chi_i^2(1) = |G|,$$

and each $\chi_i(1) \mid |G|$ (see [3]). If p is the smallest prime dividing |G| we see that

$$|G| \geq [G : G'] + (k(G) - [G : G'])p^2$$

and hence

$$k(G) \leq \frac{|G|}{p^2} + [G : G'](1 - 1/p^2).$$

In particular, if G is a non-abelian simple group, then G = G', p = 2 and $k(G) \leq 1/4(|G| + 3)$. The possibility that one can do much better than k(G) < const.|G| for non-abelian simple groups G is seen by the following argument: Let H < G be any proper subgroup of G. Then $H_G = \bigcap_{g \in G} H^g = \{1\}$. Since G acts by right multplication on the right cosets of H in G, we know that $|G| \leq [G : H]!$. Using the inequalities $s! < s^s$ and $r^r < |G|$ when $r = \frac{\ln|G|}{\ln\ln|G|}$, we have

$$\min_{H<G} [G : H] > \ln |G|/\ln \ln |G|. \qquad (2)$$

On the other hand, if |C(g)| is the order of the centralizer of $g \in G$, |C(g)| is a class invariant and $|C(g)| \cdot |Class(g)| = |G|$, so $k(G) = \frac{1}{|G|} \sum_{g \in G} |C(g)|$ for every group G. Thus

$$k(G) - 1 = \frac{1}{|G|} \sum_{g \in G-\{1\}} |C(g)| < \mathrm{Max}_{g \in G-\{1\}} |C(g)|,$$

i.e., $k(G) \leq \mathrm{Max}_{g \in G-\{1\}} |C(g)|$. When G is simple, this maximum is < |G| and we can use $|H| = \mathrm{Max}_{g \in G-\{1\}} |C(g)|$ in (2), now obtaining

$$k(G) < \ln \ln |G| \frac{|G|}{\ln|G|}.$$

Corollary 1 gives an improvement on this bound:

Corollary 1. *If* G *is a finite non-abelian simple group then* $k(G) < \frac{8|G|}{\ln|G|\cdot\ln\ln|G|}$.

Proof. For $|G|$ = either 60, 168 or 360 the inequality is easy to check. So assume that $|G| \geq 504$. Using the theorem and $\sum_{\chi \in Irr(G)} \chi^2(1) = |G|$, we obtain

$$1 + (k-1)\frac{9}{71}\ln|G|\ln\ln|G| \leq |G|.$$

Consequently $k < c|G|/\ln|G|\ln\ln|G|$ for any c satisfying

$$c - 71/9 > (\ln|G|\ln\ln|G| - 71/9)/|G|.$$

For $x \geq 504$, $f(x) = (\ln x \ln\ln x - 71/9)/x$ is a decreasing function of x, and $f(504) < 1/9$. Thus we may choose $c = 8$.

Comment. As to lower bounds for k(G), note that $k(G) > \log_2\log_2|G|$ for each group G (see [4]), although this is almost certainly too low. For example (in [1]) the first author proved that there exists a sequence $\varepsilon_n \to 0$ such that for almost all $n \leq x$, as $x \to \infty$, every group G of order n satisfies $k(G) > n^{1-\varepsilon_n}$, since G is shown to contain a correspondingly large (normal) cyclic subgroup. P Erdös remarked (in [1]) that here one may use $\varepsilon_n = f(n)\ln\ln n/\ln n$ where $f(n) \to \infty$ arbitrarily slowly, in which case $k(G) > \frac{n}{(\ln n)^{f(n)}}$ for almost all n, and G of order n. Of course none of these groups G are non-abelian simple groups, for which no lower bound better than $\log_2\log_2|G|$ has been found.

Suppose G is any group of even order $|G|$ containing exactly m involutions, and let $\mu(G) := \min\{[G:H] \mid H < G\}$. In their fundamental paper [2, p.570], Brauer and Fowler proved that if $\mu > 2$ then $\mu < \frac{n(n+2)}{2}$, where $n := \frac{|G|}{m}$. Since we (now) know that each non-abelian simple group G has even order, this gives the bound $m < \frac{|G|}{[(2\mu+1)^{1/2} - 1]}$. Using inequality (2) and the fact that $(2\mu + 1)^{1/2} - 1 \geq \mu^{1/2}$ for $\mu \geq 4$, one obtains the estimate $m < \frac{|G|}{(\ln|G|/\ln\ln|G|)^{1/2}}$. A better bound results from Corollary 1, as follows.

Corollary 2. *If* G *is a finite non-abelian simple group, and* m *the number of involutions in* G, *then* $m < \frac{\sqrt{8}\,|G|}{(\ln|G|\ln\ln|G|)^{1/2}}$.

Proof. We know from Theorem (2J) of [2] that $m(m+1) \leq |G|(k_1-1)$, where k_1 is the number of real classes in G. Thus $m^2 < |G|k(G)$, and the desired conclusion follows using Corollary 1.

Finally, we note that from Corollary (2I) of [2], applied to non-abelian simple groups G, we have

$$|G| < \left[\frac{n(n+1)}{2}\right]! \qquad (3)$$

where again $n = \frac{|G|}{m}$ and [x] is the greatest integer $\leq x$. Using Corollary 2 we can substantially improve inequality (3), as follows.

Corollary 3. *Let* G *be a finite non-abelian simple group. Then*

$$|G|^{\frac{\ln\ln|G|}{c}} < \left[\frac{n(n+1)}{2}\right]! \quad (c \sim 5.49).$$

Proof. If ln ln |G| ≤ c then the corollary follows from (3). So we may assume that ln ln |G| > c, and we shall prove that

$$|G|^{\frac{\ln\ln|G|}{c}} < \left(\frac{n}{3}\right)^{n^2} < \left[\frac{n(n+1)}{2}\right]! \qquad (4)$$

If $n \leq 7/2$, then by (3) $|G| < 7!$. Since all simple groups of order less than 7! satisfy $n > 7/2$, we may assume that $n > 7/2$. Using

$$s! > \sqrt{2\pi s}\left(\frac{s}{e}\right)^s$$

and

$$\left[\frac{n(n+1)}{2}\right] > \frac{(n+2)(n-1)}{2},$$

we obtain

$$\left[\frac{n(n+1)}{2}\right]! > \left(\frac{n(n+1)}{2e}\right)^{\frac{(n+2)(n-1)}{2}}.$$

Since $7/2 < n$, $(2e)^{1+1/n} < 9$ so

$$\left(\frac{n(n+1)}{2e}\right)^{\frac{(n+2)(n-1)}{2}} > \left(\frac{n}{3}\right)^{n^2}$$

and the right-hand inequality in (4) follows. By Corollary 2, $\left(\frac{\ln|G|\ln\ln|G|}{72}\right)^{1/2} < \frac{n}{3}$. Thus the left-hand inequality in (4) will follow from (taking logs)

$$\frac{\ln|G|\ln\ln|G|}{c} < \frac{\ln|G|\ln\ln|G|}{8} \cdot \frac{1}{2}\ln\left(\frac{\ln|G|\ln\ln|G|}{72}\right)$$

that is $\frac{72}{c}e^{16/c} < \ln|G| \left(\frac{\ln\ln|G|}{c}\right)$. Now $\ln\ln|G| > c$ and so $\ln|G| > e^c > \frac{72}{c}e^{16/c}$ for $c \geq 5.49$, and we are finished.

Acknowledgement

This article was written during the second author's visit at the University of Hawaii. He would like to thank the Department of Mathematics at the University of Hawaii for the invitation, and for their help and hospitality during his stay in Honolulu.

References

1. E A Bertram, On large cyclic subgroups of finite groups, *Proc. Amer. Math. Soc.* **56** (1976), 63-66.
2. R Brauer & K A Fowler, On groups of even order, *Ann. of Math.* (2) **62** (1955), 565-583.
3. L Dornhoff, *Group Representation Theory, Part A* (Marcel-Dekker, Inc., New York, 1971).
4. P Erdös & P Turán, On some problems of a statistical group-theory, IV, *Acta. Math. Acad. Sci. Hungar.* **19** (1968), 413-435.
5. W Feit & J G Thompson, Groups which have a faithful representation of degree less than $\frac{1}{2}$(p-1), *Pacific J. Math.* **1** (1961), 1257-1262.
6. J B Rosser & L Schoenfeld, Approximate formulas for some functions of prime numbers, *Illinois J. Math.* **6** (1962), 64-94.

FINITE PRESENTABILITY AND HEISENBERG REPRESENTATIONS

C J B BROOKES

Corpus Christi College, Cambridge, CB2 1RH

1. Introduction

Over the last decade there has been much interest in a geometric invariant Σ defined for a finitely generated group G. It takes the form of an open subset of an (n-1)-sphere S(G) where n is the torsion-free rank of the abelianisation of G. Here S(G) may be interpreted as the orbit space

$$(\mathrm{Hom}(G,\mathbf{R}) - \{0\})/\mathbf{R}_+$$

where the positive reals $\mathbf{R}_+$ act on the n-dimensional real vector space Hom(G,**R**) via scalar multiplication. Σ was originally coined by Bieri and Strebel for metabelian groups in [2]; for such a group G it encapsulates precisely whether G is finitely presentable or not. The decisive question is whether

$$S(G) = \Sigma \cup -\Sigma$$

where $-\Sigma$ is the image of Σ under the antipodal map. More recently in [1] Bieri, Neumann and Strebel have shown more generally that for a finitely generated group containing no non-abelian free subgroups the same condition is required for finite presentability, but that the condition is not sufficient to ensure it. Moreover they demonstrated that Σ also furnishes precise information as to which subgroups containing G' are finitely generated. For example G' itself is finitely generated if and only if $\Sigma = S(G)$.

Σ may be defined in various ways. The original method was to consider the monoid

$$G_\chi = \chi^{-1}([0,\infty)) = \{g \in G : \chi(g) \geq 0\}$$

for each point $[\chi]$ of S(G), and define

$$\Sigma = \{[\chi] \in S(G) : G' \text{ is finitely generated over some finitely generated submonoid of } G_\chi\}.$$

There is an alternative characterisation due to Brown concerning the existence of certain generalized HNN-extensions or, equivalently, the possible actions of G on **R**-trees with abelian length function - see [3] for details. Furthermore there is an approach using differential forms given a suitable context. For a fundamental group G of a compact connected smooth n-manifold Y, one can interpret Hom(G,**R**) as the de Rham cohomology group $H^1(Y;\mathbf{R})$ and so the points of S(G) may be represented by closed differential 1-forms on Y. Thurston defined in [11] a subset of S(G) arising from nowhere vanishing closed 1-forms on Y and in [1] this subset was shown to be the same as Σ for fundamental groups of 3-manifolds Y not involving counterexamples to the Poincaré conjecture - see [1] for a precise statement of this. Furthermore Levitt developed in [6] the relationship between Σ and the existence of certain 1-forms on manifolds of higher dimension; for example he characterized $\Sigma \cup -\Sigma$ using 'complete' 1-forms. As a non-vanishing 1-form may be used to produce a foliation of codimension one on a manifold, this approach may be viewed as a part of foliation theory.

Since writing the first version of this paper I have become acquainted with the work of the French school. The theses of Sikorav [10] and Meigniez [7] for example pursue the foliation theme providing characterisations and generalisations of Σ. The latter thesis begins the study of homomorphisms of G to Lie groups other than **R** by looking at maps to the affine group $GA_+(1,\mathbf{R})$ and to the universal cover $\overline{PSL(2,\mathbf{R})}$ of PSL(2,**R**). These Lie groups together with **R** are precisely the ones which describe orientation-preserving diffeomorphisms of the real line, the maps being regarded as holonomy representations for certain foliations of codimension one.

The aim here too is to float the idea of how to generalise using maps of G to Lie groups other than **R**. The source of inspiration though is representation theory rather foliation theory. Suppose for example that there exists a field k equipped with a non-Archimedean valuation

$$v : k \to \mathbf{R} \cup \{\infty\}$$

so that G has an image in which the image of G' is the subgroup k_0 of the additive group k and the induced conjugation action of G on k_0 is given by multiplication in the field k via a map $\theta : G \to k^*$. Let $\chi = v \circ \theta$. If G' were finitely generated over

$$G_{-\chi} = \{g \in G : \chi(g) \le 0\}$$

then k_0 would be bounded in value under v. On the other hand, if χ is non-zero, there is some g in G with $\chi(g) > 0$ and so repeated application of g to a non-zero element of k_0 would give something of arbitrarily large value in k_0. Thus $[-\chi]$

cannot lie in Σ when such an image of G exists. The field k here was a one-dimensional representation space arising from the conjugation action of G on G'.

As a generalisation one might consider representations arising from the conjugation action of G on other terms of the lower central series. This leads us towards the representation theory of nilpotent groups. Our goal is a geometric theory based on differential 1-forms and this is precisely the nature of the 'orbit method' used to describe unitary representations of Lie groups.

Left invariant differential 1-forms on a Lie group H correspond to elements of the cotangent space at the identity or, in other words, to elements of the dual space L* to the Lie algebra L of H. The Lie groups acts on L and L* via the adjoint and co-adjoint representations respectively. Kirillov showed in [4] (or see his book [5] for a fuller picture) that the co-adjoint orbits parametrise the unitary dual of a connected simply connected nilpotent Lie group. Given the earlier concentration on non-Archimedean fields it is perhaps worth noting an almost identical theory due to Moore for p-adic nilpotent Lie groups [8]. Co-adjoint orbit spaces are not the exclusive preserve of the unitary theory; they are a standard tool in the theory of algebraic group representations for example.

Rather than considering group homomorphisms $\chi : G \to \mathbf{R}$ we shall be interested in maps

$$\chi_f : G \xrightarrow{\theta} H \xrightarrow{\log} L \xrightarrow{f} \mathbf{R}$$

where H is a real nilpotent Lie group with associated Lie algebra L, θ is a homomorphism and $f \in L^*$, the dual space to L. In pursuit of universality one might consider all such χ_f factoring through a nilpotent Lie group of class at most c and observe that the real completion of $G/\gamma_{c+1}(G)$ is a universal object in the sense that all such χ_f factor through it.

Here though, in order to provide a simple illustration, we shall fix a map θ from G to the real Heisenberg group H consisting of the upper 3×3 matrices $\begin{pmatrix} 1 & \mathbf{R} & \mathbf{R} \\ 0 & 1 & \mathbf{R} \\ 0 & 0 & 1 \end{pmatrix}$ with image $H_0 = \begin{pmatrix} 1 & \mathbf{Z} & \mathbf{Z} \\ 0 & 1 & \mathbf{Z} \\ 0 & 0 & 1 \end{pmatrix}$. In Section 2 the co-adjoint structure of H is reviewed and in Section 3 a subset Σ'_θ of

$$\Omega = (L^* - \{0\})/(\mathbf{R}_+ \times H)$$

is defined, where H is acting co-adjointly and $\mathbf{R}_+$ is doing so via scalar multiplication. Note that by fixing the image of θ to be H_0 we have ensured a 1-1 correspondence between the maps χ_f and elements f of L*. The orbit space takes the form of the disjoint union of a circle and two antipodal points. The following result provides a hint that generalisations of Σ along these lines might be productive.

Theorem. *Let* G *be a finitely presentable group with no non-abelian free subgroups, and suppose there is an epimorphism* $\theta : G \to H_Q$. *Then*

$$\Omega = \Sigma'_\theta \cup -\Sigma'_\theta$$

where $-\Sigma'_\theta$ *is the image of* Σ'_θ *under the antipodal map.*

The intersection of Σ'_θ with the circle is deducible from the Bieri-Neumann-Strebel invariant of G, but whether either of the two extra points lies in Σ'_θ is perhaps providing extra information. Unfortunately I have not as yet computed an example of a group for which the circle lies entirely in $\Sigma'_\theta \cup -\Sigma'_\theta$ while neither of the two points does so.

The proof of the theorem appears in Section 4 and in the final section variants of the definition of Σ'_θ are discussed in the light of the requirements of the proof.

Acknowledgements

I should like to express my gratitude to Ralph Strebel who was indefatiguable during his visit to Cambridge to lecture to the Easter 1988 finite presentability working party sponsored by the Cambridge Philosophical Society. It is through him that I came to appreciate the subtleties of Σ. Thanks are also due to the Science and Engineering Research Council for their financial support in the form of an Advanced Fellowship.

2. The Heisenberg group

Let $h(r,s,t) = \begin{pmatrix} 1 & r & t \\ 0 & 1 & s \\ 0 & 0 & 1 \end{pmatrix} \in H$. Set $x = h(1,0,0)$, $y = h(0,1,0)$ and $z = h(0,0,1)$. Then $h(r,s,t) = y^s x^r z^t$.

The Lie algebra L associated with H is $\begin{pmatrix} 0 & \mathbf{R} & \mathbf{R} \\ 0 & 0 & \mathbf{R} \\ 0 & 0 & 0 \end{pmatrix}$ and we write $\ell(r,s,t)$ for the element $\begin{pmatrix} 0 & r & t \\ 0 & 0 & s \\ 0 & 0 & 0 \end{pmatrix}$. The exponential map $\exp : L \to H$ maps $\ell(r,s,t)$ to

$$1 + \ell(r,s,t) + \tfrac{1}{2}(\ell(r,s,t))^2 = h(r,s,t+\tfrac{1}{2}rs),$$

and the logarithmic map $\log : H \to L$ maps $h(r,s,t)$ to

$$\ell(r,s,t) - \frac{1}{2}(\ell(r,s,t))^2 = \ell(r,s,t - \frac{1}{2}rs).$$

Define $f_{a,b,c} \in L^*$ by $f_{a,b,c}(\ell(r,s,t)) = ar+bs+ct$. Thus $f_{a,b,c} \circ \log : H \to \mathbf{R}$ maps $h(r,s,t) \mapsto ar+bs+c(t - \frac{1}{2}rs)$.

The adjoint action of H on L is via matrix conjugation. The usual convention here is for H to act on the left and so the conjugate of ℓ under h is $h\ell h^{-1}$ for $h \in H$ and $\ell \in L$. This fits in with the fact that a group is taken to act on the left on a Cayley complex associated with a presentation, a construction that will be used in Section 4 in the proof of the theorem. Unfortunately the typical algebraist believes in conjugation on the right and the reader of the literature on Σ is expected to be ambidextrous. A switch of side often introduces an application of the antipodal map.

The co-adjoint action of H on L is such that the effect of h on $f_{a,b,c}$ produces ${}^h f_{a,b,c} : L \to \mathbf{R}$ sending ℓ to $f_{a,b,c}(h^{-1}.\ell) = f_{a,b,c}(h^{-1}\ell h)$. This means that ${}^h f_{a,b,c} = f_{a+cs,b-cr,c}$ where $h = h(r,s,t)$. Thus if $c = 0$ then $f_{a,b,c}$ is fixed under H. Otherwise, when $c = c_0 \neq 0$, the effect of H is to produce translations in the plane $c = c_0$; the orbit of H is the whole plane and the orbits of H_0, the discrete group $\begin{pmatrix} 1 & \mathbf{Z} & \mathbf{Z} \\ 0 & 1 & \mathbf{Z} \\ 0 & 0 & 1 \end{pmatrix}$, are lattices in the plane.

Define

$$L_{a,b,c} = \{\ell \in L:\ f_{a,b,c}(\ell) \geq 0\}$$

and

$$H_{a,b,c} = \{h \in H : \log(h) \in L_{a,b,c}\}.$$

Thus $L_{a,b,c}$ is the closed halfspace given by the inequality $ar+bs+ct \geq 0$, and

$$H_{a,b,c} = \{h(r,s,t) : ar+bs+c(t - \frac{1}{2}rs) \geq 0\}.$$

In Section 3 it will be seen that we are especially interested in $H_{0,0,1}$ and its translates. To work out $H_{a,b,c}h$, rather than calculating from first principles, one might use the Baker-Campbell-Hausdorff formula which in this context says

$$\log(h_1h_2) = \log h_1 + \log h_2 + \frac{1}{2}[\log h_1, \log h_2]$$

where [,] is the Lie bracket in L. An easy calculation shows that

$$H_{0,0,1}h(r_1,s_1,t_1) = \{h(r,s,t) : (t-t_1) - \tfrac{1}{2}(r-r_1)(s+s_1) \geq 0\}$$

and that

$$H_{0,0,-1}h(r_1,s_1,t_1) = \{h(r,s,t) : (t-t_1) - \tfrac{1}{2}(r-r_1)(s+s_1) \leq 0\}.$$

3. Definition of Σ_θ'

We are given a map $\theta : G \to H$ with $\theta(G) = H_0$ and for each $f \in L^*$ there corresponds a map $\chi_f : G \to \mathbf{R}$ given by $\chi_f = f \circ \log \circ \theta$. Set

$$G_f = \{g \in G : \chi_f(g) \geq 0\}$$

and $K = \ker \theta$. Our invariant is going to be a quotient of

$$X_\theta = \{f \in L^* : K = {}^{TG_f}\langle S\rangle \text{ for some finite sets } S \text{ and } T\}.$$

Here ${}^{TG_f}\langle S\rangle$ denotes the group generated by all elements of the form ${}^{tg}s$ with $t \in T$, $g \in G_f$ and $s \in S$. Conjugation has been written on the left to fit with the convention adopted; for example ${}^g s$ means gsg^{-1}.

Lemma 1. $f \in X_\theta$ *if and only if* $\lambda f \in X_\theta$ *for* $\lambda \in \mathbf{R}_+$.

Proof. Immediate from the fact that $G_f = G_{\lambda f}$.

Lemma 2. $f \in X_\theta$ *if and only if* ${}^h f \in X_\theta$ *for* $h \in H_0$.

Proof. This hinges on the relationship ${}^h f(\ell) = f(h^{-1}\ell h)$. Thus ${}^h f(\ell) \geq 0$ if and only if $\ell \in hL_f h^{-1}$, where $L_f = \{\ell \in L : f(\ell) \geq 0\}$. Since K is normal in G we have $K = {}^{gTG_f}\langle S\rangle$ for suitable S and T if $f \in X_\theta$. This may be rephrased as

$$K = {}^{T'gG_f g^{-1}}\langle S'\rangle$$

where $T' = {}^g T$ and $S' = {}^g S$. Using the above with $h = \theta(g)$ we deduce that ${}^h f \in X_\theta$.

These two lemmas tell us that $\mathbf{R}_+ \times H_0$ acts on X_θ and so we might consider the orbit space $X_\theta/(\mathbf{R}_+ \times H_0)$. In fact by making use of our knowledge of the H_0-orbits on L we can do a little better.

Lemma 3. *If* $f \in L^*-X_\theta$ *and* $f' \in L^*$ *lies in the convex hull of the* H_0*-orbit of* f, *then* $f' \in L^*-X_\theta$.

Proof. Suppose that $0 \neq f' = \sum_{i=1}^{n} \lambda_i{}^{h_i}f$ where $h_i \in H_0$ and $0 \leq \lambda_i \leq 1$ with $\sum_{i=1}^{n} \lambda_i = 1$. If ${}^{h_i}f(\ell) < 0$ for some $\ell \in L$ and for all h_i then $f'(\ell) < 0$. Thus

$$L_{f'} \subseteq \bigcup_{i=1}^{n} h_i L_f h_i^{-1}.$$

For f' to lie in X_θ we would need finite sets S' and T' with $K = {}^{T'G_{f'}}\langle S' \rangle$. But then we would have $K = {}^{TG_f}\langle S \rangle$ for finite sets $T = T'T_0$ and $S = {}^{T_0^{-1}}S'$ where $T_0 = \{g_1,\ldots,g_h\}$ for some g_i with $\theta(g_i) = h_i$.

In Section 2 the co-adjoint H_0-orbits were found to be either fixed points under H or lattices in a plane, the plane being a single H-orbit. The convex hull of such a lattice is the plane itself. Thus in both cases the convex hull of an H_0-orbit is actually an H-orbit. From Lemma 3 we therefore know $L^* - X_\theta$, and hence X_θ, to be invariant under H itself. We can therefore define

$$\Sigma'_\theta = X_\theta/(R_+ \times H) \subseteq L^*/(R_+ \times H).$$

4. The proof of the theorem

Our aim is to show that, when G is finitely presentable, contains no non-abelian free subgroups, and has H_0 as an epimorphic image under a map θ, the invariant Σ'_θ defined in Section 3 satisfies $\Sigma'_\theta \cup -\Sigma'_\theta = L^*/(R_+ \times H)$.

The proof is a minor variation on the original argument of Bieri and Strebel to be found in Section 4 of [2]. The idea is to let G act on a certain quotient complex Γ of the Cayley complex associated with a suitable presentation of G, to express Γ as the union of two connected subcomplexes with connected intersection and to use the Seifert-van Kampen theorem to show that the fundamental group of one of the subcomplexes must map epimorphically to the fundamental group of Γ. The crux of the matter is to define the subcomplexes so that their intersection is connected, and this proves to be possible because of a certain property of a Cayley graph of H_0.

One reason for concentrating on the Heisenberg group is that we need only look at the two extra antipodal points in $L^*/(\mathbf{R}_+ \times H)$. Any $f_{a,b,0} \in L^*$ gives rise to $\chi_{a,b,0} : G \to \mathbf{R}$ which factors through $G/\theta^{-1}(Z_0)$ where $Z_0 = \begin{pmatrix} 1 & 0 & \mathbf{Z} \\ 0 & 1 & 0 \\ 0 & 0 & 1 \end{pmatrix}$, the centre of H_0. The circle

$$S^1 = \{f_{a,b,0} \in L^* : a,b \in \mathbf{R}\}/\mathbf{R}_+$$

may be identified with $S(G,\theta^{-1}(Z_0)) \subseteq S(G)$, in the terminology of [1], and the fact that

$$(\Sigma'_\theta \cap S^1) \cup (-\Sigma'_\theta \cap S^1) = S^1$$

may be deduced from Theorem C of [1]. We shall therefore just consider the representatives $f_{0,0,1}$ and $f_{0,0,-1}$ of the two remaining points and define two subcomplexes associated with them.

We shall work with a presentation of G designed to make the corresponding Cayley complex easy to understand. Since G is finitely presentable we can find one of the form $\langle X;R \rangle$ in which the generating set X and the finite set of relations R are inverse-closed and $X = X_1 \cup X_2$ with $X_1 = \{g_1^{\pm 1}, g_2^{\pm 1}, g_3^{\pm 1}\}$ where $\theta(g_1) = x$, $\theta(g_2) = y$ and $\theta(g_3) = z$, and X_2 a finite generating set of $K = \ker \theta$ as a G-operator group. The 2-dimensional combinatorial Cayley complex $\tilde{\Gamma}$ has sets G, G×X and G×R as the sets of 0-cells, 1-cells and 2-cells respectively. The source and terminus of $(g,x_1) \in G \times X$ are g and gx_1, and the boundary of the 2-cell $(g,r) \in G \times R$ is the edge path

$$(g,x_1)(gx_1,x_2)\ldots(gx_1x_2\ldots x_{n-1},x_n)$$

where $r = x_1x_2\ldots x_n$ with $x_i \in X$. The 1-cells (g,x_1) and (gx_1,x_1^{-1}) are taken to be inverses of each other.

The group G acts on the left of $\tilde{\Gamma}$ via multiplication in G. The complex we shall consider is $\Gamma = \tilde{\Gamma}/K$. This has path group K and is acted on by $H_0 \cong G/K$ on the left. The 0-cells may be identified with elements of H_0 and the 1-skeleton may be thought of as the union of the Cayley graph of H_0 associated with generators $x^{\pm 1}$, $y^{\pm 1}$ and $z^{\pm 1}$ with a set of loops at each point. We shall be looking at the full subcomplex Δ_{s_0} generated by vertices $\{h(r,s,t) \in H_0 : s = s_0\}$. Remembering that $h(r,s,t) = y^s x^r z^t$ and that x and z commute, the geometric realisation of the 1-skeleton of Δ_{s_0} can be taken to be a 2-dimensional grid with a bouquet of circles at each vertex.

We shall define Γ_1 and Γ_2 to be the full subcomplexes of Γ generated by the vertices in $(H_{0,0,1} \cap H_0)Y_1$and $(H_{0,0,-1} \cap H_0)Y_2$ respectively for some finite sets Y_1 and Y_2 yet to be decided. To apply the Seifert-van Kampen theorem we need that $\Gamma = \Gamma_1 \cup \Gamma_2$ and that Γ_1,Γ_2 and $\Gamma_1 \cap \Gamma_2$ are connected. All these properties are proved only by reference to the Cayley graph of H_0 associated with generators $x^{\pm 1}$, $y^{\pm 1}$ and $z^{\pm 1}$.

To show that $\Gamma = \Gamma_1 \cup \Gamma_2$ we assume Y_1 and Y_2 both contain

$$\theta(X) = \{x^{\pm 1}, y^{\pm 1}, z^{\pm 1}, 1\}$$

and the tracks

$$\{1,\theta(x_1),\theta(x_1x_2),\ldots,\theta(x_1x_2\ldots x_n)\}$$

for each relator $r = x_1x_2\ldots x_n$ in R. Each $h \in H_0$ either lies in $H_{0,0,1}$ or in $H_{0,0,-1}$ and so each 0-cell h, each 1-cell $(h,\theta(x_1))$ and the boundary

$$(h,\theta(x_1))(h\theta(x_1),\theta(x_2))\ldots(h\theta(x_1x_2\ldots x_{n-1}),\theta(x_n))$$

of each 2-cell (and hence each 2-cell itself) lie in Γ_1 or Γ_2.

The other requirements to be checked concern connectivity and here it is only necessary to look at 1-skeletons. First of all, using the obvious abbreviations of notation, observe that

$$\chi_{0,0,1}(g_2^s) = 0 = \chi_{0,0,-1}(g_2^s)$$

and hence that $g_2^s \in G_{0,0,1} \cap G_{0,0,-1}$ for $s \in \mathbf{Z}$. Therefore the 1-cell $(y^s,y) \in \Gamma_1 \cap \Gamma_2$, remembering that $1 \in Y_1 \cap Y_2$. The claim is that on picking Y_1 and Y_2 suitably we can ensure that $(\Gamma_1 \cap \Delta_{s_0})$, $(\Gamma_2 \cap \Delta_{s_0})$ and $(\Gamma_1 \cap \Delta_{s_0}) \cap (\Gamma_2 \cap \Delta_{s_0})$ are all connected for all choices of $s_0 \in \mathbf{Z}$. Since for each $s \in \mathbf{Z}$ the 1-cell (y^s,y) lies in $\Gamma_1 \cap \Gamma_2$ and therefore joins

$$(\Gamma_1 \cap \Delta_s) \cap (\Gamma_2 \cap \Delta_s)$$

to

$$(\Gamma_1 \cap \Delta_{s+1}) \cap (\Gamma_1 \cap \Delta_{s+1}),$$

we shall be able to deduce from this that Γ_1, Γ_2 and $\Gamma_1 \cap \Gamma_2$ are all connected.

As already observed, the geometric realisation of the 1-skeleton of Δ_{s_0} is a 2-dimensional grid together with bouquets of circles that do not interest us because they do not affect connectivity. In Section 2 we saw that

$$H_{0,0,1}h(r_1,s_1,t_1) = \{h(r,s,t) : (t-t_1) - \tfrac{1}{2}(r-r_1)(s+s_1) \geq 0\} \qquad (*)$$

and

$$H_{0,0,-1}h(r_1,s_1,t_1) = \{h(r,s,t) : (t-t_1) - \tfrac{1}{2}(r-r_1)(s+s_1) \leq 0\}. \qquad (**)$$

Because the coefficient of t in $(t-t_1) - \frac{1}{2}(r-r_1)(s+s_1)$ is always positive whatever the choice of $h(r_1,s_1,t_1)$, we know that $(\Gamma_1 \cap \Delta_{s_0})$ is generated by all the points lying *on or above* one of a finite set of straight lines in the plane of the grid of Δ_{s_0}. In fact the subcomplex is generated by the elements of the intersection of $\{h(r,s,t) : s = s_0\}$ with the union of finitely many sets of the form (*). Likewise $\Gamma_2 \cap \Delta_{s_0}$ is generated by the points lying *on or below* one of such a finite set of lines. Both subcomplexes are therefore connected whatever the choice of Y_1 and Y_2.

To ensure the connectivity of $(\Gamma_1 \cap \Delta_{s_0}) \cap (\Gamma_2 \cap \Delta_{s_0})$ is a little trickier and requires the use of Lemma 4.2 of [2]. We have already assumed that $1 \in Y_1$ and now we select $x^{-4} = h(-4,0,0)$ to lie in Y_2. From above

$$H_{0,0,-1}x^{-4} = \{h(r,s,t) : t - \tfrac{1}{2} rs \leq 2s\}.$$

If we define $Q = \langle x,z \rangle$ and a 'valuation' $v : Q \to \mathbf{R}$ by $x^r z^t \to t - \frac{1}{2} rs_0$, then in the notation of Lemma 4.2 of [2] $\|v\| = (\frac{1}{4} s_0^2 + 1)^{1/2}$ and $v(x^{-4}) = 2s_0$. For $s_0 > 0$, the complex $(\Gamma_1 \cap \Delta_{s_0}) \cap (\Gamma_2 \cap \Delta_{s_0})$ contains the full subcomplex Δ'_{s_0} generated by the vertices

$$\{h(r,s_0,t) : 0 \leq t - \tfrac{1}{2} rs_0 \leq 2s_0\}$$

and this subcomplex is connected according to the lemma. Since $\Gamma_1 \cap \Delta_{s_0}$ is connected, any vertex $h(r,s_0,t)$ of $\Gamma_1 \cap \Delta_{s_0}$ with $t - \frac{1}{2} rs_0 < 0$ is joined by a path in $\Gamma_1 \cap \Delta_{s_0}$ to some vertex $h(r_1,s_0,t_1)$ with

$$t_1 - \tfrac{1}{2} r_1 s_0 > 2s_0.$$

This path must pass through a vertex lying in Δ'_{s_0} since v(x) and v(y) are both less than $2s_0$. Stopping at the first such vertex gives a path from $h(r,s_0,t)$ to Δ'_{s_0} lying entirely in $\Gamma_2 \cap \Delta_{s_0}$. Together with a similar argument for vertices of $\Gamma_2 \cap \Delta_{s_0}$ with $t - \frac{1}{2} rs_0 > 2s_0$, this suffices to show that $(\Gamma_1 \cap \Delta_{s_0}) \cap (\Gamma_2 \cap \Delta_{s_0})$ is connected for $s_0 > 0$ when $1 \in Y_1$ and $x^{-4} \in Y_2$. Similarly, for $s_0 < 0$, we should assume $1 \in Y_1$ and $x^4 \in Y_2$, and if $s_0 = 0$ we take $1 \in Y_1$ and $z \in Y_2$.

Thus if Y_1 and Y_2 contain suitable elements we have $\Gamma = \Gamma_1 \cup \Gamma_2$ with Γ_1, Γ_2 and $\Gamma_1 \cap \Gamma_2$ connected. The Seifert-van Kampen theorem tells us that $K = \Pi_1(\Gamma)$ is of the form $K_1 *_{K_{12}} K_2$ where $K_i = i_* \Pi_1(\Gamma_i)$ and $K_{12} = i_* \Pi_1(\Gamma_1 \cap \Gamma_2)$ where the i_* are maps of path groups (here denoted by $\Pi_1(\,)$) induced by the embeddings of Γ_1, Γ_2 and $\Gamma_1 \cap \Gamma_2$ in Γ. Since K does not contain non-abelian free subgroups we deduce that either K_1 or K_2 has index at most 2 in K. As usual, by choosing Y_1 or Y_2 to be a little larger, we can actually ensure that either K_1 or K_2 is exactly K. But K_1 is a subgroup of ${}^{G_{0,0,1}Y'_1}\langle X_2\rangle$ where Y'_1 is a (finite) set of representatives of $\theta^{-1}(Y_1)$. This is because

$$G_{0,0,1}Y'_1 = \theta^{-1}((H_{0,0,1} \cap H_0)Y_1).$$

Thus if $K = K_1$ we have that $f_{0,0,1} \in X_\theta$, and similarly if $K = K_2$ then $f_{0,0,-1} \in X_\theta$. The proof is therefore complete.

5. Remarks

(1) The condition used in the definition of X_θ is only an algebraic approximation to the statement that K_1, the image of the fundamental group of some suitable subcomplex Γ_1 of Γ, is equal to K. At the end of the proof of the theorem it was observed that K_1 is a subgroup of ${}^{G_{0,0,1}Y'_1}\langle X_2\rangle$. At this point there is a loss of information because it seems difficult to produce a precise set of generators for K_1. The geometric invariant $X_\theta/\mathbf{R}_+ \times H$ is therefore more akin to Σ'_A as defined in [1] and so it seems sensible to use the notation Σ'_θ here, in preference to Σ_θ. The problems in producing a generating set for K_1 are analogous to those discussed by Bieri, Neumann and Strebel when considering whether to use a finitely generated submonoid of G_χ rather than G_χ itself in the definition of Σ.

(2) The condition used in the definition of X_θ may be restated to say that $f \in X_\theta$ exactly when $K = {}^A\langle S\rangle$ where $A = \bigcup_{i=1}^{n} G_{f_i}$ for some representatives f_i of the H_0-orbit of f, and for some finite set S. This condition could be strengthened

to stipulate how many representatives are to be allowed. For example the case n=1 would insist that $K = {}^{G_{f_1}}\langle S\rangle$ for some element f_1 in the H_0-orbit of f. In fact by conjugating K by a suitable element of G we may assume f_1 to be f. Inspection of the proof of the theorem shows that we only needed right translates of $H_{0,0,1}$ and $H_{0,0,-1}$ and so the proof gives that either $f_{0,0,1}$ or $f_{0,0,-1}$ would lie in the n=1 analogue of X_θ.

On the other hand Lemma 3 relies on the fact that when $f' = \sum_{i=1}^{n} \lambda_i f_i$ where $f_i = {}^{h_i}f$ and $0 \le \lambda_i \le 1$, we have $L_{f'} \subseteq \bigcup_{i=1}^{n} h_i L_f h_i^{-1}$ and so $G_{f'} \subseteq \bigcup_{i=1}^{n} G_{f_i}$. Since any element of the H-orbit of $f_{0,0,1}$ may be expressed as a linear sum of only two H_0-conjugates we can be sure of the H-invariance of the n=2 analogue of X_θ. However it is not immediate that the n=1 analogue is invariant under H.

(3) In his thesis [9] Renz reinterpreted the defining condition of Σ in terms of the Cayley complex $\tilde{\Gamma}$, rather than the quotient $\tilde{\Gamma}/K$. The statement that the fundamental group of the subcomplex Γ_1 maps onto K is translated into the statement that the cover $\tilde{\Gamma}_1$ of Γ_1 in $\tilde{\Gamma}$ is 1-connected. To verify this of course one needs only to study the 1-skeleton, i.e. the Cayley graph, of $\tilde{\Gamma}$. Renz went on to generalise Σ to higher dimensions.

Meigniez's thesis [7] also adopts this approach in his consideration of maps of G to the Lie groups **R**, $GA_+(1,\mathbf{R})$ and $\overline{PSL(2,\mathbf{R})}$. He studies the connected components of the Cayley subgraphs of G generated by vertices with value no less than c, say, for $c \in \mathbf{R}$, under certain maps $\chi : G \to \mathbf{R}$ defined by using the action of G on the real line.

One might also pursue the same theme in the context of this paper by considering, for each $c \in \mathbf{R}$, the connected components of the Cayley subgraphs generated by vertices g with $\chi_{f(g)} \ge c$. The quotient of one of these Cayley subgraphs would be the 1-skeleton of an analogue of the subcomplex Γ_1 as used in the proof of the theorem. One might define an analogue of X_θ consisting of these $f \in L^*$ for which there is a choice of c giving a connected Cayley subgraph. To establish an analogue of the theorem one would wish to show that the quotient of the subgraph generated by vertices g with $a \le \chi_{f(g)} \le b$ is connected for some choice of a and b, in order to deduce the connectivity of the intersection of the analogues of Γ_1 and Γ_2. Unfortunately such quotients of subgraphs appear not to be connected.

References.

1. R Bieri, W D Neumann & R Strebel, A geometric invariant of discrete groups, *Invent. Math.* **90** (1987), 451-477.

2. R Bieri & R Strebel, Valuations and finitely presented groups, *Proc. London Math. Soc.* (3) **41** (1980), 439-464.

3. K S Brown, Trees, valuations and the Bieri-Neumann-Strebel invariant, *Invent. Math.* **90** (1987), 479-504.

4. A A Kirillov, Unitary representations of nilpotent Lie groups, *Uspekhi Mat. Nauk* **106** (1962), 57-110.

5. A A Kirillov, *Elements of the theory of representations* (Springer, Berlin, 1976).

6. G Levitt, 1-formes fermées singulières et groupe fondamental, *Invent. Math.* **88** (1987), 635-667.

7. G Meigniez, *Actions de groupes sur la droite et feuilletages de dimension un* (Thése, Lyon 1988).

8. C C Moore, Decomposition of unitary representations defined by discrete subgroups of nilpotent groups, *Ann. of Math.* **82** (1965), 146-182.

9. B Renz, *Geometrische Invarianten und Endlichkeitseigenschaften von Gruppen* (Dissertation, Frankfurt am Main 1988).

10. J C Sikorav, *Homologie de Novikov attachée à une classe de cohomologie réelle de degré un* (Thése, Orsay 1987).

11. W P Thurston, A norm on the homology of 3-manifolds, *Mem. Amer. Math. Soc.* **339** (1986).

ON NILPOTENT GROUPS ACTING ON KLEIN SURFACES

EMILIO BUJALANCE

Universidad a Distancia (UNED), 28040 Madrid, Spain

GRZEGORZ GROMADZKI

Instytut Matematyki WSP, 85-064 Bydgoszcz, Poland

1. Introduction

It is well known that a group of automorphisms of a compact Klein surface X of algebraic genus $q \geq 2$ has at most 84(q-1) elements [15]. It is also known that this bound is attained as well as it cannot be attained for infinitely many values of q ([8], [9], [10], [17], [18], [29], [31]) but the problem of the classification of those q for which this is so is far away from being solved.

If X has nonempty boundary then the bound in question can be considerably sharpened. It was C L May who started to study groups of automorphisms of such surfaces. Using rather deep analytical machinery he showed in [21] that the bound in this case is 12(q-1) and then in [13], [22], [24], [25], [27], that it is attained for infinitely many values of q (see also [11], [12], [30]). For obvious reasons bordered Klein surfaces of algebraic genus q having a group of automorphisms of order 12(q-1) were named Klein surfaces with *maximal symmetry*. The problem of classifying such surfaces up to dianalytical equivalence and corresponding groups of automorphisms up to isomorphisms seem to be very difficult. However certain pieces of it seem to be more approachable. Recently for example May [27] has classified topological types of bordered Klein surfaces with maximal symmetry and supersoluble group of automorphisms showing in particular that such a surface of algebraic genus $q \geq 3$ must be orientable. In [6] the bound for the order of a supersoluble group of automorphisms of bordered nonorientable Klein surfaces is found and the corresponding surfaces are classified up to topological type. For the Riemann surfaces this problem was completely solved in [14] (see also [35]).

Recently a number of results concerning the above problems for nilpotent groups have appeared. Although some of them have a certain topological flavour they essentially concern certain purely group theoretical problems. The results make a fairly complete picture, some are still to be printed and the aim of this article is to give a short survey of them.

2. Methods

By a Klein surface we shall mean a compact topological surface equipped with dianalytic structure. This object seems to be already known to Klein [16], however the modern study of it is due to Alling and Greenleaf [1]. Every Klein surface X of algebraic genus $q \geq 2$ can be represented as a quotient space D/Γ, where D is the upper complex half plane with hyperbolic structure and Γ is a certain discrete group of isometries of D including those whose reverse orientation - reflections and glide-reflections *(NEC group)* [28]. Moreover Γ can be assumed to be a *surface NEC group* i.e. an NEC group with signature

$$(g; \pm; [-], \{(-), \overset{k}{\ldots}, (-)\})$$

(see below). The number g corresponds to the topological genus, the sign to the orientability and k is just the number of boundary components of X. If X is not a Riemann surface i.e. if the sign is "-" or $k > 0$ then the number $q = \alpha g+k-1$ is called the *algebraic genus of* X (or of Γ) and it is the genus of the field extension $\mathcal{M}(X)/\mathbf{R}$, where $\mathcal{M}(X)$ is the field of meromorphic functions on X and is the same as the topological genus of the canonical double Riemann cover of X. In the case of Riemann surfaces the algebraic genus q equals the topological one g.

On the other hand having a surface so represented, a finite group G can be faithfully represented as a group of its automorphisms if and only if there exists an NEC group Λ such that $G \cong \Lambda/\Gamma$.

The algebraic structure of Λ is determined by its signature [19], [32]

$$(g; \pm; [m_1,...,m_r], \{(n_{i1},...,n_{is_i})_{i=1,...,k}\}). \tag{2.1}$$

A group Λ with signature (2.1) has the presentation with the following generators

(i) x_i, $i = 1,...,r$,

(ii) c_{ij}, $i = 1,...,k$, $j = 0,...,s_i$,

(iii) e_i, $i = 1,...,k$,

(iv) a_i,b_i, $i = 1,...,g$ (if the sign is "+")

d_i, $i = 1,...,g$ (if the sign is "-")

subject to the relations

(i) $x_i^{m_i} = 1, \quad i = 1,...,r,$

(ii) $c_{is_i} = e_i^{-1} c_{i0} e_i, \quad i = 1,...,k,$

(iii) $c_{i,j-1}^2 = c_{i,j}^2 = (c_{i,j-1}c_{ij})^{n_{ij}} = 1, \quad i = 1,...,k; \quad j = 1,...,s_i,$

(iv) $x_1 \ldots x_r e_1 \ldots e_k a_1 b_1 a_1^{-1} b_1^{-1} \ldots a_g b_g a_g^{-1} b_g^{-1} = 1$ if the sign is "+",

$x_1 \ldots x_r e_1 \ldots e_k d_1^2 \ldots d_g^2 = 1$ if the sign is "-".

We shall refer to the above generators as the *canonical generators* of Λ.

Every NEC group has a *fundamental region* whose hyperbolic area depends only on the algebraic structure of the group and for a group with signature (2.1) is given by

$$\mu(\Lambda) = 2\pi(\alpha g + k - 2 + \sum_{i=1}^{r} (1 - 1/m_i) + \sum_{i=1}^{k} (1 - 1/n_{ij})/2). \qquad (2.2)$$

Moreover if Λ' is a subgroup of an NEC group Λ of finite index then Λ' is also an NEC group and the following Riemann Hurwitz index formula holds

$$[\Lambda:\Lambda'] = \mu(\Lambda')/\mu(\Lambda). \qquad (2.3)$$

Now since a surface group Γ of algebraic genus q has area $2\alpha\pi(q-1)$, where $\alpha = 2$ if Γ is Fuchsian and $\alpha = 1$ otherwise, we see that the problem of finding an upper bound for the order of a group of automorphisms (lying in a certain family of finite groups $\mathfrak{F}$) of a Klein surface of algebraic genus $q \geq 2$ is equivalent to the problem of finding a lower bound for the area of NEC groups Λ for which there exist surface groups Γ with $\Lambda/\Gamma \in \mathfrak{F}$. We shall refer to such a factor as an $\mathfrak{F}$ *surface kernel (bordered or unbordered, orientable or nonorientable) factor* according to whether $k > 0$ or $k = 0$ and the sign is "+" or "-" in Γ. Having established such a bound, the problem of its realization is nothing else than the study of possible orders of factors Λ/Γ, where Λ is an NEC group with minimal possible area. The problem of the topological classification of corresponding Klein surfaces is equivalent to the problem of the study of possible signatures that a group Γ may have. As we shall see these problems are equivalent to purely group theoretical ones.

In recent years a combinatorial method (based mainly on a surgery of fundamental regions of NEC groups) of determining algebraic structure of normal subgroups Λ' of an NEC group Λ as a function of the orders of the images in Λ/Λ' of the canonical generators of Λ and certain of their products was developed in a series of articles [2], [3], [6], [7]. The shortage of space does not give us the possibility of demonstrating how these results work. However it will be pointed out in the text where we use it, referring to it as c-*method* or as c-*arguments* (combinatorial).

3. Riemann surfaces

This is the most classical case. Such a surface can be represented as D/Γ, where Γ is a Fuchsian surface group i.e. an NEC group with signature (g; +; [-], {-}).

Recently Zomorrodian [33] has shown that if Λ is a *Fuchsian group* (NEC group without elements reversing orientation of D) and Λ/Γ is nilpotent then $\mu(\Lambda) \geq \pi/4$ and the bound in this case is attained just for the group Λ with signature (0; +; [2,4,8], {-}). It is also easy to show that G can be represented as such a factor if and only if it is generated by two elements a and b of orders 2 and 4 whose product has order 8. (All these facts can be easily deduced using c-arguments.) We see that such a factor must be a 2-group of order 2^n ($n \geq 4$). On the other hand a group G with presentation $< x,y \mid x^2, y^4, xyxy^5 >$ is such a factor, say Λ/Γ of order 16. Taking a k-term of the 2-Frattini series of Γ,

$$\Gamma_1^{(2)} = \Gamma^2[\Gamma,\Gamma],\ \Gamma_{n+1}^{(2)} = (\Gamma_n^{(2)})_1^{(2)}$$

for suitable k, Zomorrodian showed that there exists a surface Fuchsian subgroup Γ' of Γ normal in Λ such that the order of a quotient Λ/Γ' is greater than or equal to that arbitrarily presumed. Next using the fact that a 2-group has a normal series with cyclic factors of order 2 he easily argues that there exists such a factor of order 2^n for arbitrary $n \geq 4$. We have presented the above construction, that has obviously to be attributed to Macbeath [18], because it and its different variants are very useful and in fact basic in realization problems. We shall refer to it in what follows as F-*arguments*. In that way Zomorrodian obtained in [33] the following result.

A nilpotent group of orientation preserving automorphisms of a compact Riemann surface of genus $q \geq 2$ *has no more than* $16(q-1)$ *elements. Moreover the bound is attained if and only if* $q = 2^n+1$ *for some* $n \geq 4$.

Maximal nilpotent groups of automorphisms of compact Riemann surfaces turn out to be 2-groups. This led Zomorrodian to consider the analogous problem of finding a bound for the order of a p-group (p an odd prime) of automorphisms of a compact Riemann surface. Using considerations similar to the preceding ones it is very easy to obtain his main results [34].

A p-group of automorphisms of a compact Riemann surface of genus $g \geq 2$ *has at most*

$9(g-1)$ *if* $p = 3$,

$2p(g-1)/(p-3)$ *if* $p \geq 5$,

elements.

Moreover the necessary and sufficient condition for the existence of a Riemann surface for which the above bound is attained is that

$g-1 = 3^n$, *for some* $n \geq 2$, *if* $p = 3$ *and*

$g-1 = (p-3)p^n/2$, *for some* $n \geq 0$ *if* $p \geq 5$.

4. Bordered orientable Klein surfaces

Let $G = \Lambda/\Gamma$ be a bordered surface kernel factor. Then using c-arguments one can show (c.f. [6]) that $\mu(\Lambda) \geq \pi/6$ and that the bound is attained only for an NEC group with signature (0; +; [-], {(2,2,2,3)}). Therefore from the Riemann Hurwitz formula we obtain that a bordered Klein surface of algebraic genus $q \geq 2$ has at most $12(q-1)$ elements. (The original proof of this result due to May [21] uses analytical arguments.) Moreover a finite group G can be represented as such a factor if and only if it can be generated by three elements a, b and c of order 2 such that ab and ac have orders 2 and 3 respectively. In particular we see that such a factor cannot be nilpotent. Recently May [26] has shown that if such a factor group $G = \Lambda/\Gamma$ is nilpotent then $\mu(\Lambda) \geq \pi/4$ and if $\mu(\Lambda) = \pi/4$ then Λ is an NEC group with signature (0; +; [-], {(2,2,2,4)}). A necessary and sufficient condition for G to be represented as a bordered surface kernel factor of Λ is that G can be generated by three elements a, b and c of order 2 such that ab and ac have orders 2 and 4 respectively. Furthermore if bc has order q then Γ can be assumed to have $k = |G|/2q$ boundary components. It is also possible to show that Γ can be assumed to be nonorientable if and only if ab and ac generate the whole group G [6]. In particular we see that a nilpotent group of automorphisms of a bordered Klein surface of algebraic genus $q \geq 2$ has at most $8(q-1)$ elements [26]. We also see that the problem of topological classification of bordered Klein surfaces admitting nilpotent groups of automorphisms of maximal possible order is equivalent to the following purely group theoretic problem.

Problem. Given $q = 2^{\alpha}$ ($\alpha \geq 0$) find possible orders of 2-groups admitting three generators a, b and c of order 2 such that ab, ac and bc have orders 2, 4 and q respectively and then determine when G can be generated by ab and ac.

All above results can be proved using c-arguments [6].

Example (May [26]). Given $n \geq 0$ there exists an orientable compact Klein surface of algebraic genus $q = 2^n+1$ with q boundary components and nilpotent group of automorphisms of order $8(q-1)$.

(1) *There are only two topological types of Klein surfaces of algebraic genus* $q=2$ *with a nilpotent group of automorphisms of order* 8: *a torus with one hole and a real projective plane with two holes* (May [26]).

Recently it was proved in [6] that

(2) *If* X *is a Klein surface of algebraic genus* $q \geq 3$ *with* k *boundary components having nilpotent group of automorphisms of order* $8(q-1)$ *then*

(i) X *is orientable* [4],

(ii) $q-1 = 2^n$, *for some integer* n,

(iii) $k = 2^m$ *for some integer* m *such that* $1 \leq m \leq \begin{cases} 2 & \text{if } n-1, \\ n & \text{if } n \neq 1. \end{cases}$

(3) *Conversely given integers* n *and* m *such that* $1 < m \leq \begin{cases} 2 & \text{if } n=1, \\ n-1 & \text{if } n \neq 1, \end{cases}$

there exists an orientable Klein surface of algebraic genus $q = 2^n+1$ *with* $k = 2^m$ *boundary components and with nilpotent group of automorphisms having order* $8(q-1)$.

Notice that the above result togther with the example of May gives a complete topological classification of Klein surfaces admitting a nilpotent group of automorphisms of maximal possible order.

The most difficult to prove among necessary conditions (2) is (i). It was proved in [4] and it will be discussed in the coming section.

In the proof of (3) we start with a group with presentation

$$< x,y,w \mid x^2, y^2, w^{2^r}, [x,y], w^x = w^{2^{r-1}+1}, w^y = w^{-1} >,$$

where $r = n-m+2$. This group being a semidirect product of the cyclic group of order 2^r by the Klein four group has order 2^{r+2} and clearly $A = x$, $B = y$ and $C = yw$ have order 2, they generate G and AB, AC and BC have orders 2, 4 and 2^r respectively. Applying c-arguments we argue that G is isomorphic to the factor Λ/Γ, where Λ is an NEC group with signature $(0; +; [-], \{(2,4,2^r)\})$ and Γ is a surface Fuchsian group with signature (g; ±; [-], {-}). Applying F-arguments and the fact that a 2-group has a normal series with cyclic factors of order 2 (c.f. Section 3) we argue that a group $\tilde{G}$ of order 2^{n+3} being a factor of such a form exists. Using c-arguments again we argue that it is equivalent to the existence of three generators $\tilde{A}$, $\tilde{B}$ and $\tilde{C}$ of $\tilde{G}$ of order 2 such that $\tilde{A}\tilde{B}$, $\tilde{A}\tilde{C}$ and $\tilde{B}\tilde{C}$ have orders 2,4 and $q = 2^r$ respectively, which completes the proof.

Nilpotent groups of automorphisms of bordered Klein surfaces of maximal possible order turn out to be 2-groups. This led May in [26] to ask for the bound of the order of p-groups of automorphisms of such surfaces. In [4] we proved the following.

A p-*group* (p *an odd prime) of automorphisms of an orientable compact Klein surface of algebraic genus* $q \geq 2$ *has at most* $\frac{p}{p-2}(q-1)$ *elements.*

Now given $n \geq 1$ let N(n) be the smallest integer such that a group of order $p^{N(n)}$ generated by two elements of order p whose product has order p^n exists. Notice that such a number exists for any $n \geq 1$. In fact for $n \geq 2$ one can show that a k-term of the p-Frattini series $\Gamma_k^{(p)}$ of a Fuchsian group Γ with signature

$$(0; +; [p,p,p^n], \{-\})$$

is a Fuchsian surface group for k large enough (e.g. k = n see [20]). The case n = 1 is obvious; N(1) = 1.

In [6] the following result is proved classifying topologically orientable Klein surfaces with boundary admitting a p-group of automorphisms of maximal possible order up to knowledge of the sequence $\{N(n)\}_{n \in \mathbb{N}}$.

The necessary and sufficient condition for the existence of an orientable Klein surface of algebraic genus $q \geq 2$ *with* k *boundary components and* p-*group of automorphisms of order* $\frac{p}{p-2}(q-1)$ *is the existence of* $n \geq 0$ *and* $s \geq 0$ *such that:*

(a) $q = (p-2)p^n+1$,

(b) $N(s) \leq n+1$,

(c) $k = p^{n-s+1}$.

We finish this section with a purely group theoretic problem.

Problem. Calculate N(n) for $n \geq 2$. (During the Conference Avinoam Mann (Jerusalem) showed us that N(2) = p+1. However it seems that for arbitrary n this is a rather difficult problem.)

5. Bordered nonorientable Klein surfaces

May who started the study of groups of automorphisms of such surfaces in [26] was able to give only one example of such a surface: a real projective plane with 2 holes (c.f. Section 4). This led him to ask in a former version of [26] if the bound $8(q-1)$ for this class of surfaces can be improved for $q \geq 3$ and to conjecture in the revised one that the answer to this question is no. Recently we have shown in [4] a group theoretical result.

Lemma. *A finite nilpotent group* G *of order greater than* 8 *cannot be generated by three elements* A, B *and* C, *of order* 2 *such that* AB *and* BC *generate the whole group* G *and* AB, BC *and* AC *have orders* 2, k *and* m *respectively, where* k *and* m *are greater than* 2.

The importance of this lemma lies in the fact that it shows that the approach, suggested in [26, Problem 2, p.292], for finding nonorientable Klein surfaces of algebraic genus $q \geq 3$ with nilpotent group of automorphisms of order $8(q-1)$ fails.

Using this lemma we proved in [4] by c-arguments that:

A nilpotent group of automorphisms of a nonorientable bordered Klein surface of algebraic genus $q \geq 3$ *has at most* $4q$ *elements. Moreover this bound is attained if and only if* q *is a power of* 2, *the corresponding group of automorphisms is the dihedral group and the corresponding Klein surface has* q *boundary components.*

We see that $4q = 8(q-1)$ for q=2 only. Thus our theorem gives a negative answer to the conjecture of May.

In the case of p-groups we proved in [4] that

A p-*group* ($p \neq 2$) *of automorphisms of a nonorientable bordered Klein surface of algebraic genus* $q \geq 2$ *has at most* $\frac{p}{p-1}(q-1)$ *elements.*

Finally in [6] it is proved that

The necessary and sufficient condition for the existence of a nonorientable Klein surface of algebraic genus $q \geq 2$ *with* k *boundary components and* p-*group of automorphisms of order* $\frac{p}{p-1}(q-1)$ *is the existence of* $n \geq 0$ *such that:*

(a) $q = (p-1)p^n + 1$,

(b) $k = \begin{cases} 1 \text{ or } p & \text{when } n = 0 \\ p^r \text{ for some } r \text{ in the range } 0 \leq r \leq n, & \text{when } n > 0. \end{cases}$

6. Nonorientable Riemann surfaces

Using the lemma of the previous section we prove in [4] using c-arguments that:

A nilpotent group of automorphisms of a compact nonorientable Klein surface without boundary of algebraic genus $q \geq 2$ *has at most* $8(q-1)$ *elements. Moreover this bound is attained for* $q = 2$.

The proof of this result consists in showing that if Λ/Γ is a nilpotent nonorientable unbordered surface kernel factor then $\mu(\Lambda) \geq \pi/4$ and the bound in this case may be attained only for NEC groups with signatures:

$$(0; +; [2], \{(2,4)\}), (0; +; [4], \{(4)\}),$$

$$(0; +; [-], \{(4,4,4)\}), (0; +; [-], \{(2,2,2,4)\}).$$

However we are only able to produce one of these factors (for a group with the last signature). It would be interesting to know whether this case is like the nonorientable bordered one (see previous section). In this respect we pose the following:

Problem. Determine whether or not the bound $8(q-1)$ for the order of a group of automorphisms of a nonorientable Riemann surface of algebraic genus q can be improved for $q \geq 3$. Of course if the answer to this question is positive another interesting problem is to find the sharp bound and to classify topologically corresponding Klein surfaces.

Remark. A group of automorphisms of a nonorientable Klein surface without boundary can be viewed as a group of orientation-preserving automorphisms of a Riemann surface of the same algebraic genus. Thus we see that the bound $16(q-1)$ for a nilpotent group of automorphisms of a compact Riemann surface is not attained in the nonorientable case. It is worth noting that this is in marked contrast with the corresponding results for the maximal groups of automorphisms of compact Klein surfaces for which the bound $84(q-1)$ (in the case of surfaces without boundary) and $12(q-1)$ (in the case with boundary) is attained both in the orientable and in the nonorientable cases (see for example [31] and [13] respectively).

In the case of p-groups one can prove the following (c.f. [4]):

A p*-group of automorphisms of a nonorientable Klein surface without boundary of algebraic genus* $q \geq 2$, *has at most* $\frac{p}{p-2}(q-1)$ *elements. Moreover given a positive integer* n *there exists a* p*-group of order* p^n *acting as a group of automorphisms on a nonorientable Klein surface without boundary of algebraic genus* $q = (p-2)p^{n-1} + 1$.

Acknowledgements

The authors thanks to J M R Sanjurjo for helpful comments. The first author was partially supported by CICYT. The paper was written while the author was visiting professor at Complutense University in Madrid.

References

1. N L Alling & N Freenleaf, *Foundations of the theory of Klein surfaces* (Lecture Notes in Math. **219**, Springer Verlag 1971).
2. E Bujalance, Proper periods of normal NEC subgroup swith even index, *Rev. Mat. Hispano-Americana* (4) **41** (1981), 121-127.
3. E Bujalance, Normal subgroup sof NEC groups, *Math. Z.* **178** (1981), 331-341.
4. E Bujalance & G Gromadzki, On nilpotent groups of automorphisms of compact Klein surfaces, *Proc. Amer. Math. Soc.*, to appear.
5. E Bujalance & E Martinez, A remark on NEC groups of surfaces with boundary, *Bull. London Math. Soc.* **21**(3) (1988), 263-266.
6. E Bujalance, J J Etayo Gordejuela, J M Gamboa & G Gromadzki, *Automorphism groups of compact bordered Klein surfaces - a combinatorial approach* (Lecture Notes in Math., Springer Verlag), to appear.
7. J A Bujalance, Normal subgroups of even index in an NEC group, *Arch. Math.* **49** (1987), 470-478,
8. J Cohen, On Hurwitz extensions of $PSL_2(7)$, *Math. Proc. Cambridge Philos. Soc.* **86** (1979), 395-400.
9. M D E Conder, Generators for alternating and symmetric groups, *J. London Math. Soc.* **22** (1980), 75-86.
10. M D E Conder, The genus of compact Riemann surfaces with maximal automorphism group, *J. Algebra* **108** (1987), 204-247.
11. J J Etayo Gordejuela, Klein surfaces with maximal symmetry and their groups of automorphisms, *Math. Ann.* *268* (1984), 533-538.
12. J J Etayo Gordejuela & C Perez-Chirinos, Bordered and unbordered Klein surfaces with maximal symmetry, *J. Pure Appl. Algebra* **42** (1986), 29-35.
13. N Greenleaf & C L May, Bordered Klein surfaces with maximal symmetry, *Trans. Amer. Math. Soc.* **274** (1982), 265-283.
14. G Gromadzki & C Maclachlan, Supersoluble groups of automorphisms of compact Riemann surfaces, *Glasgow Math. J.* **31** (1989), 321-327.
15. A Hurwitz, Uber algebraische Gebilde mit eindeutigen Transformationen in sich, *Math. Ann.* **41** (1893), 402-442.
16. F Klein, *Uber Riemanns Theorie der algegraischen Funktionen and ihrer Integrale* (Teubner, Leipzig 1882).

17. J Lehner & M Newman, On Riemann surfaces with maximal automorphism groups, *Glasgow Math. J.* **8** (1967), 102-112.

18. A M Macbeath, On a theorem of Hurwitz, *Proc. Glasgow Math. Assoc.* **5** (1961), 90-96.

19. A M Macbeath, The classification of non-euclidean plane crystallographic groups, *Canad. J. Math.* **19** (1967), 1192-1205.

20. A M Macbeath, Residual nilpotency of Fuchsian groups, *Illinois J. Math.* **28**(2) (1984), 299-311.

21. C L May, Automorphisms of compact Klein surfaces with boundary, *Pacific J. Math.* **59** (1975), 199-210.

22. C L May, Large automorphisms groups of compact Klein surfaces with boundary, I, *Glasgow Math. J.* **18** (1977), 1-10.

23. C L May, A bound for the number of automorphisms of a compact Klein surface with boundary, *Proc. Amer. Math. Soc.* **63** (1977), 273-280.

24. C L May, Maximal symmetry and fully wound coverings, *Proc. Amer. Math. Soc.* **79** (1980), 23-31.

25. C L May, A family of M*-groups, *Canad. J. Math.* **37** (1986), 1094-1109.

26. C L May, Nilpotent automorphism groups of bordered Klein surfaces, *Proc. Amer. Math. Soc.* **101**(2) (1987), 287-292.

27. C L May, Supersolvable M*-groups, *Glasgow Math. J.* **30** (1988), 31-40.

28. R Preston, *Projective structures and fundamental domains on compact Klein surfaces* (Ph.D. Thesis, Univ. of Texas 1975).

29. C H Sah, Groups related to compact Riemann surfaces, *Acta Math.* **123** (1969), 13-42.

30. D Singerman, PSL(2,q) as an image of the extended modular group with applications to group actions on surfaces, *Proc. Edinburgh Math. Soc.* **30** (1987), 143-151.

31. D Singerman, Automorphisms of compact non-orientable Riemann surfaces, *Glasgow Math. J.* **12** (1971), 50-59.

32. M C Wilkie, On non-euclidean crystallographic groups, *Math. Z.* **91** (1966), 87-102.

33. R Zomorrodian, Nilpotent automorphism groups of Riemann surfaces, *Trans. Amer. Math. Soc.* **288** (1985), 241-255.

34. R Zomorrodian, Classification of p-groups of automorphisms of Riemann surfaces and their lower central series, *Glasgow Math. J.* **29** (1987), 237-244.

35. R Zomorrodian, Supersoluble groups of automorphism groups of Riemann surfaces, preprint 1988.

SOME ALGORITHMS FOR POLYCYCLIC GROUPS

FRANK B CANNONITO

University of California, Irvine, CA 92717, USA

1. Introduction

The algorithmic theory of some solvable groups appears now to be rather complete. In the near future two rather large papers will appear which will give complete details of the algorithms for polycyclic-by-finite groups devised by me together with Gilbert Baumslag, Derek J S Robinson and Dan Segal [BCRS]. In addition, Dan Segal will give additional algorithms for automorphisms of these groups which will conclude with a positive solution for the isomorphism problem [DS]. Another development that bears watching is the actual implementation of some of these algorithms for high speed computers. Indeed, Charles Sims has almost got the so-called "polycyclic quotient algorithm" devised by Gil Baumslag, myself and Chuck Miller [BCM IV] running; something that very few of us dreamed could happed (see [CS]). So for the first time a practical theory of computing in group theory with some real depth seems likely. No one can tell where this could lead.

The property of polycyclic-by-finite groups that facilitates this development is that they are *coherent*. That is, they and all their subgroups are finitely presented. Another important feature is that these groups have an easily recognized (finite) presentation. Indeed, if one writes down a so-called polycyclic presentation (see below), which inevitably presents a solvable group, then the polycyclic quotient algorithm referred to above enables you to effectively decide if the group presented is polycyclic [BCM IV]. This means that there is an abundance of presentations of polycyclic groups at hand to try the new algorithms on and you know which these presentations are.

In fact the whole development goes over to the slightly wider class of polycyclic-by-finite groups. These groups are those which have a normal polycyclic subgroup of finite index. Indeed, using the algorithm in [BCRS] which presents the solvable radical, you can effectively get your hands on the "almost polycyclic" part of one of these groups and proceed to apply the algorithmic theory already developed to it; the finite "top" piece really offers no great obstacle to continuing the algorithm on to the whole group.

Since this subject will have a speedy dissemination in the publications cited above, I will refrain from surveying what can be done and, rather, I will give some variants of the algorithms concocted for finding the maximal abelian normal subgroup and the centre of a polycyclic group. These variants, which were devised together with Gil Baumslag, may turn out to have some features that make them more palatable to those engaged in future efforts to implement these algorithms on computing machines, for reasons known only to the experts and not to me. In describing the algorithms below, I will make essential use of Neil Maxwell's discovery which survives in the fragment of his Ph.D. thesis and is embodied in the following:

Theorem 1.1 (N Maxwell). *Let* G *be a polycyclic group given by an explicit finite presentation. Then there exists an algorithm to construct a finite presentation of* Fitt(G), *the Fitting subgroup of* G.

(A different proof is given for this theorem in [BCRS].)

It is only fitting that this discussion start with the consequences of this constructive aspect of the Fitting subgroup and that I be writing about this in a paper connected with my lecture at 'Groups - St Andrews 1989', since it was at 'Groups - St Andrews 1985' that Dan Segal first told me about Maxwell's result and this was the key that enabled Gil Baumslag and me to present the centre of an explicitly given polycyclic group. This algorithm had eluded us prior to hearing about Maxwell's result.

2. Basic properties of polycyclic groups

I will assume the following widely known facts about polycyclic groups. The reader who wishes details for these assumptions should consult the book [PG]. First, polycyclic groups are Hopfian. That is, any epimorphism of a polycyclic group onto itself is an isomorphism. Second, polycyclic groups have solvable generalized word problem which means that, with respect to any fixed presentation for the group, there is an algorithm which when presented with a finite list $w_1,...,w_{k+1}$ of words on the generators of the presentation, decides if $w_{k+1} \in \langle w_1,...,w_k \rangle$. Specializing to the case $k = 0$, we see that polycyclic groups have solvable word problem.

The next definition was given in [BCM I]. It is central to the entire development of the algorithmic theory of polycyclic-by-finite groups.

Definition 2.1. A *polycyclic presentation* of a group G is a finite presentation with generators $x_1,\dots,x_n$ and relations

$$x_i^{x_j} = v_{ij}(x_1,\dots,x_{j-1}),$$

$$1 < i < j < n,$$

$$x_i^{x_j^{-1}} = v_{ij}'(x_1,\dots,x_{j-1}),$$

and

$$x_i^{\theta_j} = u_i(x_1,\dots,x_{i-1}), \qquad 1 < i < n,$$

where v_{ij}, v_{ij}', u_i are certain words and $1 < \theta_i < \infty$, the third relation being vacuous if $\theta_i = \infty$.

Let $G_i = < x_1,\dots,x_i >$. Then there is a series

$$1 = G_0 \triangleleft G_1 \triangleleft \dots \triangleleft G_n = G,$$

and G_i/G_{i-1} is cyclic of order dividing θ_i. Call the presentation *consistent* if it induces a presentation of each G_i in $x_1,\dots,x_i$ (in which case $|G_i:G_{i-1}| = \theta_i$). Clearly, a group has a consistent polycyclic presentation if and only if it is polycyclic. (This definition is a minor variant of the one used in [BCM I] where the term "honest" is used for "consistent".)

Then it is not hard to show, as is done in [BMC I] that there exists a uniform algorithm to decide if a polycyclic presentation is consistent. Accordingly, we may as well assume that any presentation we are working with for a polycyclic group actually is consistent. Since I will only be discussing polycyclic groups this will suffice; however, an analogous situation prevails for polycyclic-by-finite groups, as noted above.

3. The algorithms

We begin by quoting the following:

Theorem 3.1 [BCM I]. *Let* G *be a polycyclic group given by an explicit finite presentation. Then there exists an algorithm which, when presented with finitely many words* $w_1,\dots,w_k$ *on the generators of* G, *gives a finite presentation for*

$$H = < w_1,\dots,w_k >.$$

The proof is by induction on the number of factors in a consistent polycyclic presentation for G, the so-called "polycyclic length" of G, and involves the Euclidean algorithm.

Another well-known result needed is given by:

Lemma 3.2. *Let* $U = U(n,\mathbf{Z})$ *be the group of* $n \times n$ *lower unitriangular matrices over* $\mathbf{Z}$. *Then* U *has an easily described consistent polycyclic presentation which can be explicity given.*

Corollary 3.3. *If* H *is explicitly given as a finitely generated subgroup of* $U(n,\mathbf{Z})$ *then there is an algorithm to explicitly present* H.

This leads to the first:

Theorem 3.4. *Suppose* A *is a finitely generated torsion-free abelian subgroup of a finitely generated nilpotent group* G *which we may assume is given by a consistent polycyclic presentation. Then we can effectively decide if* A *is self centralizing (i.e. if* $C_G(A) = A$).

Proof. Map $G/A \to \mathrm{Aut}(A)$ by conjugation;

$$gA \mapsto \{x \mapsto g^{-1}xg\ (x \in A)\}.$$

Call this map ϕ. Then A is self centralizing if and only if ϕ is monic since $\ker\phi$ centralizes A. Identify Aut(A) with $Gl(n,\mathbf{Z})$. Suppose G is generated by $x_1,\ldots,x_p$. Then G/A is generated by $x_1A,\ldots,x_pA$. We present G/A by adding to the presentation for G finitely many relators which kill A. We next present

$$\phi(G/A) = < \phi(x_1A),\ldots,\phi(x_pA) >.$$

To do this notice each x_iA acts nilpotently on A. That is,

$$A > [A,G] > [A,G,G] > \ldots > [A,_nG] = 1,$$

for some n. This means that we can choose a basis for A at the outset so that $\phi(G/A)$ is contained in $U(n,\mathbf{Z})$ and that, hence, $\phi(G/A)$ has an easily computed finite presentation.

To check that ϕ is monic we take the relators holding in the presentation for $\phi(G/A)$ and rewrite them in terms of the x_iA. Then, using the solution for the word problem in G/A, we check to see if these are 1. If so, A is self centralizing. If not, A is not self centralizing.

Corollary 3.5. *Suppose* A *is any finite generated abelian normal subgroup of the finitely generated nilpotent group* G. *There exists an algorithm to decide if* A *is self centralizing.*

Proof. We decompose $A = \mathbf{Z}^n \times F$, where F is a finite abelian group. Here Aut F is finite and the map $\phi : G/A \to \text{Aut } A$ can be further analysed as follows. Write

$$A = \{(a,f) : a \in \mathbf{Z}^n, f \in F\}.$$

Put

$$X = \{\alpha \in \text{Aut } A : \alpha(a,f) = (a',f), \text{ for some } a' \in \mathbf{Z}^n\}.$$

In other words, X consists of those automorphisms of A that fix F. Then $X \cong Gl(n,\mathbf{Z})$. We let

$$Y = \{\alpha \in \text{Aut } A : \alpha(a,f) = (a,f') \text{ for some } f' \in F\}.$$

Here Y is the group of automorphisms of A which fix $\mathbf{Z}^n$.

We note

$$\text{Aut } A = XY,$$

for if $\delta \in \text{Aut } A$ and $\delta(a,f) = (a',f')$, then the map $a \longmapsto a'$ defines an element α of $Gl(n,\mathbf{Z})$; that is, an element of X. Now

$$\alpha^{-1}\delta(a,f) = (a,f'),$$

and so $\beta = \alpha^{-1}\delta$ is an element of Y. Hence $\delta = \alpha\beta$ and Aut A = XY. Note $X \lhd \text{Aut } A$ since X is the kernel of the map $\alpha \longmapsto \alpha|F$ (we use the fact that F is characteristic in A here). Thus

$$\text{Aut } A = X \rtimes Y.$$

Suppose $\beta \in Y$. Then $\beta(1,f) = (1,f')$ and so with $\beta' = \beta|F$ we see the map $f \longmapsto f'$ is an automorphism of F. If we consider $\beta^{-1}\beta'$ we see that $\beta^{-1}\beta'$ acts as the identity on F and also as the identity on A/F and so $\beta^{-1}\beta'$ is a derivation of $\mathbf{Z}^n$ into F. Hence we interpret β' as an automorphism of A. Then since $\beta' = \beta\gamma$ where $\gamma = \beta^{-1}\beta'$ is an element of $\text{Der}(\mathbf{Z}^n,F)$ we have

$$Y = \text{Aut } F \cdot \text{Der}(\mathbf{Z}^n,F).$$

This gives

$$\text{Aut } A = Gl(n,\mathbf{Z}) \rtimes \text{Aut } F \cdot \text{Der}(\mathbf{Z}^n,F).$$

But Aut F is a finite nilpotent group and $\text{Der}(\mathbf{Z}^n,F)$ is a finite abelian group and as above we can arrange for $\phi(G/A)$ to lie in $U(n,\mathbf{Z}) \rtimes \text{Aut } F \cdot \text{Der}(\mathbf{Z}^n,F)$. Hence we can explicitly compute a presentation for $\phi(G/A)$. We now proceed as in the proof of Theorem 3.4 to decide if ϕ is monic and, hence, if A is self centralizing. This concludes the proof.

Corollary 3.6. *There exists an algorithm by means of which we can compute a presentation of a maximal abelian normal subgroup in any explicity given finitely generated nilpotent group.*

Proof. Suppose we have our hands on some abelian normal subgroup A of G. For example we may take $\gamma_c(G)$ if G is nil-c. We check to see if A is self centralizing. If so, we are done. If not, map $G/A \to \text{Aut } A$ and denote this map by ϕ. Let ker $\phi' = K/A$. Now $K \lhd G$ since K is the preimage of $K/A \lhd G/A$ and K is the centralizer of A in G.

We present K effectively and consider

$$[K/A,G/A],\ [K/A,\ G/A,\ G/A],\ldots$$

There exists a least d, which can effectively be found by using the algorithm in [BMC IV] to test for triviality, such that

$$L/A = [K/A,_{d}G/A] \neq 1,$$

and

$$[K/A,_{d+1}G/A] = 1.$$

We present L/A and choose zA in L/A with $z \notin A$. Then z is central modulo A and $z \in K$ (since by normality everything lies in K). Then $A_1 = < z,A >$ is abelian and

$$A_1 = < z,A > \lhd G.$$

(A_1 is abelian since $z \in K = C_G(A)$ and A is abelian; A_1 is normal in G because modulo A it is central in G and so the preimage of a normal subgroup.) Since $A_1 > A$ and this construction can be iterated we must eventually arrive at a subgroup A_k which is maximal normal abelian in G and this k can be found by the usual method effectively.

Corollary 3.7. *There is an algorithm whereby we can present effectively the centre of an explicitly presented finitely generated nilpotent group* G.

Proof. Let A be a maximal abelian normal subgroup of G, where G is presented on generators $x_1,\ldots,x_p$ and compute using [BCM III]

$$A_i = \text{ann}_{ZA}(x_i - 1)$$

for $i = 1,\ldots,p$. Since A is self centralizing, $\zeta A < A$ and

$$\zeta G = \bigcap_i A_i.$$

By [BCM III] this intersection can be effectively presented as a finitely presented ZA-module and, hence, ζG can be effectively presented as an abelian group.

This gives the main result.

Theorem 3.8. *There exists an algorithm to compute the centre of an explicitly presented polycyclic group* G.

Proof. Let F = Fitt(G) and A = ζF. By Theorem 1.1 and Corollary 3.7 we can effectively present F and A. Then if G is generated by $x_1,\dots,x_p$, we have $\zeta G < F$ and, hence, $\zeta G < A$. Thus

$$\zeta G = \cap A_i \quad (i = 1,\dots,p),$$

where A_i is as above. The remainder of the proof is similar.

4. Conclusion

The algorithmic theory of polycyclic-by-finite groups is quite extensive. There are some surprising results, not the least among which is the existence of an algorithm to present the Frattini subgroup, despite the highly nonconstructive character of its definition. As mentioned above this result will appear in the paper [BCRS]. Some of the most elementary aspects of polycyclic-by-finite group theory sometimes require the most strenuous efforts to produce the algorithms wanted. Thus the intersection of two finitely generated subgroups, the centralizer and the normalizer of a finitely generated subgroup all can be effectively presented, but the proofs for these results are much deeper than those presented above. I have no idea if there is any chance to program these, but given that no one in 1981 would have guessed that the polycyclic quotient algorithm had a prayer of being realized on computers I think the prudent attitude is to wait and see.

In my talk at St Andrews I discussed some possibilities of duplicating these ideas in the class of finitely generated metabelian groups. Here, there are some immediate obstacles to overcome stemming from the fact that some important subgroups such as the commutator subgroup [G,G] need not be finitely generated. Nevertheless, using ideas from [BCM II] and [BCM III] there is a way to effectively present the commutator subgroup as a finitely presented Z(G/[G,G])-module. This leads to the existence of algorithms to give a mixed presentation, part group-theoretic and part module-theoretic, for the Fitting subgroup and the centre. This will appear shortly in [BCR]. Whether or not there will be chance to realize these algorithms remains to be seen.

I wish to express my thanks to Gilbert Baumslag and Derek Robinson with whom so much of this work was created, and for all the stimulating hours we passed together during which I learned so much and had such great fun.

Literature

[BCM I] Gilbert Baumslag, F B Cannonito & C F Miller III, Infinitely generated subgroups of finitely presented groups, I, *Math. Z.* **153** (1977), 117-134.

[BCM II] Gilbert Baumslag, F B Cannonito & C F Miller III, Infinitely generated subgroups of finitely presented groups, part II, *Math. Z.* **172** (1980), 97-105.

[BCM III] Gilbert Baumslag, F B Cannonito & C F Miller III, Computable algebra and group embeddings, *J. Algebra* **69** (1981), 186-212.

[BCM IV] Gilbert Baumslag, F B Cannonito & C F Miller III, Some recognizable properties of solvable groups, *Math. Z.* **178** (1981), 289-295.

[BCR] Gilbert Baumslag, F B Cannonito & Derek J S Robinson, The algorithmic theory of metabelian groups, part I, to appear.

[BCRS] Gilbert Baumslag, F B Cannonito, D J S Robinson & Dan Segal, The algorithmic theory of polycyclic-by-finite groups, *J. Algebra*, to appear.

[CS] Charles C Sims, Implementing the Baumslag-Cannonito-Miller polycyclic quotient algorithm, *J. Symbolic Comput.*, to appear.

[DS] Dan Segal, Decision properties of polycyclic groups, to appear.

[PG] Dan Segal, *Polycyclic Groups* (Cambridge Tracts in Mathematics **82**, Cambridge Univ. Press, Cambridge 1983).

ON THE REGULARITY CONDITIONS FOR COLOURED GRAPHS

ANTONIO F COSTA

Facultad de Ciencias de la U.N.E.D., Madrid 28040, Spain

1. Coloured graphs

Let Γ be a connected graph (loops are forbidden but multiple edges are allowed). $V(\Gamma)$ will denote the set of vertices of Γ. Assume that Γ is regular of degree $h \geq 2$ and that there is a map

$$\gamma : E(\Gamma) \to \Delta_{n-1} = \{i \in \mathbf{Z}; 0 \leq i \leq h-1\}$$

such that $\gamma(e_1) \neq \gamma(e_2)$, for any two adjacent edges e_1, e_2, then the pair (Γ,γ) is an h-coloured graph and the map γ is an edge-colouration of Γ. Given an (n+1)-coloured graph (Γ,γ) it is possible to construct an n-dimensional pseudocomplex $K(\Gamma,\gamma)$ in the following way: take an n-simplex $\sigma^n(x)$ for each $x \in V(\Gamma)$ and label its vertices by Δ_n, and if $x,y \in V(\Gamma)$ are joined by an edge e with $\gamma(e) = c \in \Delta_n$ then identify the (n-1)-faces of $\sigma_n(x)$ and $\sigma_n(y)$ opposite to vertices labelled by c so that equally labelled vertices are identified together (see [FGG]). It is easy to embed (Γ,γ) in $K(\Gamma,\gamma)$ in a "coloured manner" (see [BM]) and then the coloured graphs are a natural generalization to higher dimensions of maps on surfaces. In this paper we present some different extensions of the concept of regular maps on surfaces to coloured graphs and their relations with group theory.

2. Regular coloured graphs and locally regular coloured graphs

The regular coloured graphs are introduced by A Vince in 1983, ([V]). The regularity is defined in terms of the automorphism group of the coloured graph.

If (Γ,γ) is a coloured graph, an automorphism of (Γ,γ) is an automorphism ψ of the graph Γ such that $\gamma \bullet \psi = \gamma$. The group of automorphisms of Γ is denoted by $Aut(\Gamma,\gamma)$.

Definition 2.1 (Regular coloured graph [V]). A coloured graph (Γ,γ) is *regular* if $\mathrm{Aut}(\Gamma,\gamma)$ acts transitively on V(G).

A group theoretical characterization of regular coloured graphs is given in [V] using the fact that any h-coloured graph can be considered as a Schreier coset graph, $G(W_h,H)$, where W_h is a group generated by involutions, r_i, $i \in \Delta_h$, and H is a subgroup of W_h (the points of $G(W_h,H)$ are the right cosets of W_h/H and two points u and u' are i-adjacents, $i \in \Delta_h$, if and only if $u' = ur_i$).

Proposition 2.2 ([V]). $G(W_h,H)$ *is a regular coloured graph if and only if* H *is a normal subgroup of* W_h.

In the case of regular coloured graphs (Γ,γ) with $K(\Gamma,\gamma)$ closed simply connected manifolds the following relation with Coxeter group is proved in [V]:

Proposition 2.3 ([V]). *Let* (Γ,γ) *be a regular* h-*coloured graph,* Γ *without cycles formed by two edges and assume that* $K(\Gamma,\gamma)$ *is a closed simply connected manifold. Then* (Γ,γ) *is* $G(W_h,H)$ *with* W_h/H *a finite Coxeter group.*

The notion of locally regular coloured graphs is a generalization of the concept of regular coloured graphs which is easier to check on given concrete examples.

Definition 2.4 ([C2]). An h-coloured graph (Γ,γ) is a *locally regular* h-*coloured graph* if for each $i,j \in \Delta_h$ all the $\{i,j\}$-coloured cycles in (Γ,γ) have the same number of edges.

Proposition 2.3 is a particular case of the following result:

Theorem 2.5 ([C2]). *Let* (Γ,γ) *be an* (n+1)-*locally regular coloured graph such that* $K(\Gamma,\gamma)$ *is a manifold. Then there is a regular* (n+1)-*coloured graph* $(\Gamma^{\sim},\gamma^{\sim})$ *such that* $K(\Gamma^{\sim},\gamma^{\sim})$ *is a tessellation by n-simplexes of* X, *where* X *is either* H^n, E^n *or* S^n *and there is a subgroup* G *of* $\mathrm{Aut}(\Gamma^{\sim},\gamma^{\sim})$, *such that* G *acts freely on* X, *and* $(\Gamma^{\sim},\gamma^{\sim})/G = (\Gamma,\gamma)$.

Remark 2.6. With the conditions of the last theorem, G is a normal subgroup of $\mathrm{Aut}(\Gamma^{\sim},\gamma^{\sim})$ iff (Γ,γ) is regular.

3. Semiregular coloured graphs

There are many manifolds which cannot be triangulated by a pseudocomplex of the type $K(\Gamma,\gamma)$ with (Γ,γ) a locally regular coloured graph (by Theorem 2.5 the manifold $|K(\Gamma,\gamma)|$ would admit a spherical, euclidean or hyperbolic structure and in [C2] it is proved that there are manifolds having one of these geometrical structures and not admitting "locally regular coloured triangulations"). In order to have a richer list of manifolds among the $K(\Gamma,\gamma)$ in [CG] we introduce the concept of semiregular coloured graphs inspired by the idea of the regular embeddings of coloured graphs [G].

Definition 3.1 ([CG]). An (n+1)-coloured graph (Γ,γ) is said to be *semiregular of type* $(t_0,\dots,t_n) \in \mathbf{Z}^{n+1}$ if there is a cyclic permutation $\varepsilon = (\varepsilon_0,\dots,\varepsilon_n)$ of Δ_n such that each cycle with edges coloured alternately by ε_i and ε_{i+1} has $2t_i$ edges.

In [CG] it is proved that every triangulable manifold is homeomorphic to $|K(\Gamma,\gamma)|$ where (Γ,γ) is a semiregular coloured graph (see [CG]). For the semiregular coloured graphs there is also a nice relation with group theory:

Proposition 3.2 ([CG], see also [C1]). *There is a* 1-1 *correspondence between the* (n+1)-*coloured graphs of type* $(t_0,\dots,t_n)$ *and the surface subgroups of a* 2-*dimensional crystallographic group of signature:*

$$(0; [-]; \{(t_0,\dots,t_n)\}).$$

For these types of coloured graphs there are also universal graphs $K(\Gamma^{\sim},\gamma^{\sim})_{t_0,\dots,t_n}$ (as in 2.5 for locally regular coloured graphs). The pseudocomplexes $K(\Gamma^{\sim},\gamma^{\sim})_{t_0,\dots,t_n}$ and $K(\Gamma^{\sim},\gamma^{\sim})_{t'_0,\dots,t'_n}$ are homeomorphic if and only if the two signatures

$$(0; [-]; \{(t_0,\dots,t_n)\}) \text{ and } (0; [-]; \{(t_0,\dots,t_n)\})$$

correspond both to either spherical, euclidean or hyperbolic crystallographic groups.

References

[BM] J Bracho & L Montejano, The combinatorics of coloured triangulations of manifolds, *Geom. Dedicata* **22** (1987), 303-328.

[C1] A F Costa, Coloured graphs representing manifolds and universal maps, *Geom. Dedicata* **28** (1988), 349-357.

[C2] A F Costa, Locally regular coloured graphs, preprint.

[CG] A F Costa & L Grasselli, Fuchsian groups and coloured graphs, preprint.

[FGG] M Ferri, C Gagliardi & C Grasselli, A graph-theoretical representation of P.L.-manifolds - A survey on crystallizations, *Aequationes Math.* **31** (1986), 121-141.

[G] C Gagliardi, Regular imbeddings of edge-coloured graphs, *Geom. Dedicata* **11** (1981) 397-414.

[V] A Vince, Regular combinatorial maps, *J. Combin. Theory Ser. B* **35** (1983), 256-277.

MULTIPLET CLASSIFICATION OF HIGHEST WEIGHT MODULES OVER QUANTUM UNIVERSAL ENVELOPING ALGEBRAS: THE $U_Q(SL(3,C))$ EXAMPLE

V K DOBREV

Insitute of Nuclear Research and Nuclear Energy, 1784 Sofia, Bulgaria

1. Introduction and preliminaries

1.1. The q-deformation $U_q(\mathcal{G})$ of the universal enveloping algebras $U(\mathcal{G})$ of simple Lie algebras $\mathcal{G}$ arose in the study of the algebraic aspects of quantum integrable systems [1-4]. (The definition of $U_q(\mathcal{G})$ is below in 1.2.) The algebras $U_q(\mathcal{G})$ are called also quantum groups [5] or quantum universal enveloping algebras [6,7]. They provide a powerful tool for the solving of the quantum Yang-Baxter equations. In [4b] for $\mathcal{G}$ = sl(2,C) and in [5,8] in general it was observed that the algebras $U_q(\mathcal{G})$ have the structure of a Hopf algebra. This brought additional mathematical interest in this new algebraic structure (see, e.g., [9-12]). The representations of $U_q(\mathcal{G})$ were studied in [3,7,11,12] for generic values of the deformation parameter. Actually all results from the representation theory of $\mathcal{G}$ carry over to the quantum group case. This is not so, however, if the deformation parameter q is a root of unity. There are only a few results in this case [13,14,15] mainly for $\mathcal{G}$ = sl(2,C). On the other hand it turns out that the case when q is a root of unity is also very interesting for the physical applications, e.g. in rational conformal field theories [14,16].

Thus in this paper we address the representation theory of $U_q(\mathcal{G})$ when the deformation parameter q is a root of unity. We study the induced highest weight modules (HWM) over $U_q(\mathcal{G})$ (Section 2). They all are reducible for $q^N = 1$, $N \in N+1$. Thus it is natural to use the previously developed approach of multiplet classification of HWM over (infinite-dimensional) (super-) Lie algebras [17-21]. Using this we give in Section 3, the multiplet classification of the induced HWM over $U_q(\mathcal{G})$ when $\mathcal{G}$ = sl(2,C), sl(3,C) and a complete classification of the irreducible HWM in the case $\mathcal{G}$ = sl(3,C). For illustration of our approach we also give the irreducible HWM over sl(2,C) which were obtained in [13,14,15]. It is clear that our approach can be used for other $U_q(\mathcal{G})$. Work along these lines is in progress.

1.2. Let $\mathcal{G}$ be any simple Lie algebra; then $U_q(\mathcal{G})$ is defined [5,8] as the associative algebra over C with generators $X_i^{\pm}$, H_i, $i = 1,\dots,r = \text{rank}\,\mathcal{G}$ and with relations

$$[H_i, H_j] = 0,\ [H_i, X_j^{\pm}] = \pm\, a_{ij} X_j^{\pm}, \tag{1}$$

$$[X_i^{+}, H_j^{-}] = \delta_{ij} \frac{q_i^{H_i/2} - q^{-H_i/2}}{q_i^{1/2} - q_i^{-1/2}} = \delta_{ij}\, [H_i]_{q_i},\quad q_i = q^{(\alpha_i,\, \alpha_i)/2}, \tag{2}$$

$$\sum_{k=0}^{n} (-1)^k \binom{n}{k}_{q_i} (X_i^{\pm})^k\, X_j^{\pm}\, (X_i^{\pm})^{n-k} = 0,\quad i \neq j, \tag{3}$$

where $(a_{ij}) = (2(\alpha_i, \alpha_j)/(\alpha_i, \alpha_i))$ is the Cartan matrix of $\mathcal{G}$, $(\cdot,\cdot)$ is the scalar product of the roots normalized so that for the long simple roots α we have $(\alpha, \alpha) = 2$, $n = 1 - a_{ij}$, $\binom{n}{k}_q = \frac{[n]_q!}{[k]_q!\,[n-k]_q!}$, $[m]_q! = [m]_q\,[m-1]_q \dots [1]_q$,

$$[m]_q = \frac{q^{m/2} - q^{-m/2}}{q^{1/2} - q^{-1/2}} = \frac{\text{sh}(mh/2)}{\text{sh}(h/2)} = \frac{\sin(\pi m\tau)}{\sin(\pi\tau)},\quad q = e^h = e^{2\pi i\tau}. \tag{4}$$

This definition is valid also for Kac-Moody algebras [5]. Further we shall omit the subscript q in $[m]_q$ if no confusion can arise.

For $q \to 1$, $(h \to 0)$, we recover the standard commutator relations from (1), (2) and Serre's relations from (3) in terms of the Chevalley generators H_i, $X_i^{\pm}$. We recall that H_i corresponds to the simple roots α_i of $\mathcal{G}$, and if $\beta = \Sigma_i\, n_i\, \alpha_i$, then to β corresponds $H_\beta = \Sigma_i\, n_i\, H_i$. The elements H_i span the Cartan subalgebra $\mathcal{H}$ of $\mathcal{G}$, and we shall use the standard decomposition $\mathcal{G} = \mathcal{G}^+ \oplus \mathcal{H} \oplus \mathcal{G}^-$.

In [4b] for $\mathcal{G} = sl(2,C)$ and in [5,8] in general it was observed that the algebra $U_q(\mathcal{G})$ is a Hopf algebra [22] with co-multiplication δ, co-unit ε and antipode γ defined on the generators and on the unit element of $U_q(\mathcal{G})$ as follows:

$$\delta(H_i) = H_i \otimes 1 + 1 \otimes H_i,$$

$$\delta(X_i^{\pm}) = X_i^{\pm} \otimes q_i^{-H_i/4} + q_i^{H_i/4} \otimes X_i^{\pm},\quad \delta(1) = 1 \otimes 1, \tag{5a}$$

$$\varepsilon(H_i) = \varepsilon(X_i^{\pm}) = 0,\quad \varepsilon(1) = 1, \tag{5b}$$

$$\gamma(H_i) = -H_i, \quad \gamma(X_i^{\pm}) = -q_i^{-\hat{\rho}/2}\, X_i^{\pm}\, q_i^{-\hat{\rho}/2}, \quad \gamma(1) = 1, \tag{6}$$

where $\hat{\rho} \in \mathcal{H}$ corresponds to $\rho = \frac{1}{2}\Sigma_{\alpha\in\Delta^+}\ \alpha$, Δ^+ is the set of positive roots, $\hat{\rho} = \frac{1}{2}\Sigma_{\alpha\in\Delta^+}\ H_\alpha$.

2. Reducibility and embeddings of highest weight modules

2.1. The highest weight modules V over $U_q(\mathcal{G})$ [6] are given by their highest weight $\lambda \in \mathcal{H}^*$ and highest weight vector $v_0 \in V$ such that:

$$X_i^+\, v_0 = 0, \qquad i = 1,\dots,r, \quad Hv_0 = \lambda(H)v_0, \quad H \in \mathcal{H}. \tag{7}$$

We start with the induced HWM V^λ such that

$$V^\lambda \cong U_q(\mathcal{G}) \otimes_{U_q(\mathcal{B})} v_0 \cong U_q(\mathcal{G}^-) \otimes v_0,$$

where $\mathcal{B} = \mathcal{B}^+$, $\mathcal{B}^\pm = \mathcal{H} \oplus \mathcal{G}^\pm$ are Borel subalgebras of $\mathcal{G}$. (Then the algebras $U_q(\mathcal{B}^\pm)$ with generators H_i, $X_i^\pm$ are Hopf subalgebras of $U_q(\mathcal{G})$ [6].) Thus one should expect that the representation theory of V^λ would parallel the theory of Verma modules $V(\Lambda)$ over $\mathcal{G}$. ($V(\Lambda)$ is defined as the HWM over $\mathcal{G}$ induced from the one-dimensional representations of $\mathcal{B}$.) In particular, we shall consider the irreducible HWM L_λ over $U_q(\mathcal{G})$ as factor-modules V^λ/I^λ, where I^λ is the maximal submodule of V^λ.

We recall several facts from the case q = 1 (see, e.g. [23]). The Verma module $V(\Lambda)$ over $\mathcal{G}$ is reducible iff there exists a root $\beta \in \Delta^+$ and $m \in \mathbf{N}$ such that

$$2(\Lambda + \rho,\beta) = m(\beta,\beta) \tag{8}$$

holds [24]. (In the Kac-Moody case (8) is the condition of Kac-Kazhdan [25].) If (8) holds then there exists a vector $v_s \in V^\lambda$, called a *singular vector*, such that $v_s \neq v_0$, $X_i^+\, v_s = 0$, $i = 1,\dots,r$, $Hv_s = (\Lambda(H) - m\beta(H))v_s$, $H \in \mathcal{H}$. (In the affine Kac-Moody case when $(\beta,\beta) = 0$ there are p(n) independent singular vectors for each $n \in \mathbf{N}$, $p(\cdot)$ being the partition function [26].) $U(\mathcal{G}^-)v_s$ is a proper submodule of $V(\Lambda)$ isomorphic to the Verma module $V(\Lambda - m\beta) = U(\mathcal{G}_-) \otimes v_0'$ where v_0' is the highest weight vector of $V(\Lambda - m\beta)$; the isomorphism being realized by $v_s \mapsto 1 \otimes v_0'$. The singular vector is given by

$$v_s^{\beta,m} = \mathcal{P}_m^\beta (X_1^-, ..., X_r^-) \otimes v_0, \tag{9}$$

where $\mathcal{P}_m^\beta$ is a homogeneous polynomial in its variables of degrees mn_i, where $n_i \in \mathbf{Z}_+$ come from $\beta = \Sigma\, n_i\alpha_i$, α_i - the system of simple roots [17,18]. $V(\Lambda)$ contains a unique proper maximal submodule $V'(\Lambda)$. Among the HWM with highest weight Λ there is a unique irreducible one, denoted by $L(\Lambda)$, i.e.,

$$L(\Lambda) = V(\Lambda)/V'(\Lambda). \tag{10}$$

If $V(\Lambda)$ is irreducible then $L(\Lambda) = V(\Lambda)$. Thus we further discuss $L(\Lambda)$ for which $V(\Lambda)$ is reducible. Consider $V(\Lambda)$ reducible with respect to (w.r.t.) every simple root (and thus w.r.t. all positive roots):

$$(\Lambda + \rho, \alpha_i^\vee) = m_i, \qquad i = 1, ..., r. \tag{11}$$

Then $L(\Lambda)$ is a finite-dimensional highest weight module. (In the affine Kac-Moody case, $L(\Lambda)$ with (11) holding are the so-called integrable highest weight modules [27].) An important class of the case when (11) holds are the so-called *fundamental representations* $L(\Lambda_i)$, $i = 1, ..., r$ characterized by $(\Lambda_i, \alpha_j^\vee) = \delta_{ij}$, i.e. $(\Lambda_i + \rho, \alpha_j^\vee) = 1 + \delta_{ij} = m_j(\Lambda_i)$. In the case when q is not a root of unity all facts above hold also for the representations of $U_q(\mathcal{G})$ [3,7,11,12].

In the case when q is a root of unity there are a few results [13,14,15], however, they are not in our context. We start with the case of the simple roots. Let $\beta = \alpha_j$ and we try the same expression (9) for the singular vector as in the case $q = 1$:

$$v_s = (X_j^-)^m \otimes v_0. \tag{12}$$

We obtain using (2):

$$[X_j^+, (X_j^-)^m] = \sum_{k=0}^{m-1} (X_j^-)^{m-1-k}[H_j](X_j^-)^k$$

$$= (X_j^-)^{m-1} \sum_{k=0}^{m-1} [H_j - 2k] = (X_j^-)^{m-1}[m][H_j - m + 1].$$

If v_s is a singular vector we should have

$$0 = X_j^+ v_s = [X_j^+, (X_j^-)^m] \otimes v_0 = (X_j^-)^{m-1}[m][\lambda(H_j) - m + 1] \otimes v_0. \tag{13}$$

(Note that $X_k^+ v_s = 0$, for $k \neq j$.) If $q_j = q^{(\alpha_j,\alpha_j)/2}$ is not a root of unity (13) gives just condition (8) rewritten as $\lambda(H_j) = (\lambda,\alpha_j^\vee) = m - (\rho,\alpha_j^\vee) = m - 1$, where $\beta^\vee = 2\beta/(\beta,\beta)$, $((\rho,\alpha_j^\vee) = 1$ for all $\alpha_j)$. If q_j is a root of unity, then $q_j^{N_j} = 1$, $N_j \in \mathbf{N} + 1$ for $k \in \mathbf{Z}$:

$$[kN_j]_{q_j} = \frac{q_j^{kN_j/2} - q_j^{-kN_j/2}}{q_j^{1/2} - q_j^{-1/2}} = \frac{\sin(\pi k)}{\sin(\pi/N_j)} = 0, \quad q_j = e^{2\pi i/N_j}.$$

Accordingly (13) gives that v_s from (12) is a singular vector iff

$$\text{either } [m]_{q_j} = 0, \quad \forall\lambda \in \mathcal{H}^*, \tag{14a}$$

$$\text{or } [\lambda(H_j) + 1 - m]_{q_j} = 0. \tag{14b}$$

Thus we see that for $q_j^{N_j} = 1$ the HWM V^λ is always reducible. Analogously, all vectors of the form $v_s^{k_1,\dots,k_r} = \prod_{j=1}^{r} (X_j^-)^{k_jN_j} \otimes v_0$, $k_j \in \mathbf{Z}_+$, $\sum_{j=1}^{r} k_j > 0$, are singular vectors. One reason is that actually all elements $(X_j^-)^{kN_j}$, (also $(X_j^+)^{kN_j}$), belong to the centre of $U_q(\mathcal{G})$. Thus the general form of the singular vector corresponding to $\beta \in \Delta^+$, $\beta = \Sigma\, n_j\alpha_j$, and $m = kN+n$, where N is such that $q_j = e^{2\pi i/N}$ for the shortest simple roots α_j which enter the decomposition of β, $k,n \in \mathbf{Z}_+$, $k+n \neq 0$, $n < N$, is

$$v_s^{\beta,n,k} = \prod_{j=1}^{r} (X_j^-)^{kn_jN}\, \mathcal{P}_n^\beta(X_1^-, \dots, X_r^-) \otimes v_0, \tag{15a}$$

where $\mathcal{P}_n^\beta$ is a homogeneous polynomial as in (9) and if $n \neq 0$ then the condition

$$[(\lambda + \rho)(H_\beta) - n] = 0, \qquad n \in \mathbf{N}, \qquad n < N, \tag{15b}$$

is fulfilled.

Let us say that two elements $\lambda,\lambda' \in H^*$ are *equivalent*, $\lambda \cong \lambda'$, if $\lambda-\lambda' = N\beta$, where β is any element of the dual integer root lattice, i.e., $\beta = n_1\alpha_1^\vee + \ldots n_r\alpha_r^\vee$, $n_i \in \mathbf{Z}$, $\alpha_i^\vee = 2\alpha_i/(\alpha_i,\alpha_i)$, and N is such that $q_j = e^{2\pi i/N}$ for the shortest simple roots α_j whose duals enter the decomposition of β.

It is clear that if $\lambda \cong \lambda'$, then they obey or disobey (14b), (15b) simultaneously. Thus the HWM V^λ and $V^{\lambda'}$ have the same structure and the corresponding irreducible factor-modules will be equivalent. So the irreducible HWM are described by their highest weights up to the above equivalence. Because of (15) it is also clear that the irreducible HWM of $U_q(\mathcal{G})$ are finite-dimensional. Moreover, one may work with the finite algebra setting all elements $(X_i^-)^{kN_i}$, $(X_i^+)^{kN_i}$ equal to zero. (In some explicit realizations this actually happens [15].)

2.2. We consider the question of the irreducible representations as quotients of reducible V^λ HWM in the framework of embeddings between such HWM. It is clear that the HWM $V^{\lambda'}$ is isomorphic to a submodule of V^λ if $\lambda \cong \lambda'$ and $\lambda - \lambda' = N\beta$ for β an element of the dual non-negative integer root lattice, i.e.,

$$\beta = n_1\alpha_1^\vee + \ldots n_r\alpha_r^\vee,\ n_i \in \mathbf{Z}_+.$$

Thus to account for all other embeddings it is enough to consider the singular vectors in (15) with $k = 0$, $n \in \mathbf{N}$, $n < N$. It is clear that if (15b) holds then $V^{\lambda-n\beta}$ is isomorphic to a submodule of V^λ. If (15b) holds for several pairs $(n,\beta) = (m_i,\beta_i)$, $i = 1,\ldots,k$, there are other HWM modules $V^{\lambda-m_i\beta_i}$ all of which are isomorphic to submodules of V^λ. Furthermore if (15b) holds with $\beta \in \Delta^+$ and $n \in -\mathbf{N}$ then V^λ is a submodule of $V^{\lambda+n\beta}$. Indeed, if $[(\lambda+\rho)(H_\beta) + n] = 0$ then $[(\lambda + n\beta + \rho)(H_\beta) - n] = 0$ because $\beta(H_\beta) = 2$ for all β.

What is more interesting and in contrast to the undeformed $q = 1$ case is that if V^λ has a singular vector of type (15) with $k = 0$, $n = m \in \mathbf{N}$, $m < N$ then the embedded HWM $V^{\lambda-m\beta}$ has a singular vector of type (15) with $k = 0$, $n = N-m$. The embedded HWM in $V^{\lambda-m\beta}$ is easily seen to be $V^{\lambda-N\beta}$. The latter is a submodule also of V^λ, however with a singular vector from (15) with $k = 1$, $n = 0$. The two embeddings coincide if $\beta = \alpha_i$ is a *simple* root. Before demonstrating this let us fix the usual convention that the arrows depicting the embedding maps point *to* the embedded HWM. Indeed, the first embedding is a composition of two embeddings

$$V^\lambda \to V^{\lambda-m\beta} \to V^{\lambda-N\beta};$$

correspondingly if v_0', v_0'' are the highest weight vectors of $V^{\lambda-m\beta}$, $V^{\lambda-N\beta}$, resp., we have

$$\mathcal{P}^{\beta}_{N-m}\mathcal{P}^{\beta}_{m} \otimes v_0 \mapsto \mathcal{P}^{\beta}_{N-m} \otimes v_0' \mapsto 1 \otimes v_0'';$$

the second embedding is $V^{\lambda} \to V^{\lambda-N\beta}$; under this we have

$$\prod_{j=1}^{r} (X_j^-)^{n_j N} \otimes v_0 \mapsto 1 \otimes v_0'',$$

where $\beta = \Sigma\, n_j\alpha_j$. Thus if β is not a simple root we may have embedding of one and the same module in two different ways. This is similar to the affine Kac-Moody case when β is an imaginary root (i.e., $(\beta,\beta) = 0$).

It is convenient for the description of the embedding patterns to use the notion of a multiplet [17,18] of highest weight modules. We say that a set $\mathcal{M}$ of highest weight modules forms a *multiplet* if

(1) $V \in \mathcal{M} \Rightarrow \mathcal{M} \supset \mathcal{M}_V$, where $\mathcal{M}_V$ is the set of all highest weight modules $V' \neq V$ such that either V' is isomorphic to a submodule of V or V is isomorphic to a submodule of V';

(2) $\mathcal{M}$ does not have proper subsets fulfilling (1).

It is convenient to depict a multiplet by a connected oriented graph, the vertices of which correspond to the highest weight modules and the lines between the vertices correspond to the embeddings between the modules. Note that it may happen that two multiplets $\mathcal{M}^1$ and $\mathcal{M}^2$ are depicted by one and the same graph. Then we say that $\mathcal{M}^1$ and $\mathcal{M}^2$ belong to one and the same *type of multiplets* and use parametrization to distinguish multiplets belonging to a certain type. Most often only the embeddings which are not compositions of other embeddings are depicted, since these contain all the relevant information.

3. The examples of $U_q(sl(2,C))$ and $U_q(sl(3,C))$

3.2. We start with the example of $\mathcal{G} = sl(2,C)$; $r = 1$, $X_1^{\pm} = X^{\pm}$, $H_1 = H$, $\alpha_1 = \alpha = \alpha^{\vee} = 2\rho$. We shall prove that all HWM V^{λ} belong to multiplets of one of the two types described below.

3.2.1. The multiplets of the first type are in 1-1 correspondence with those equivalent classes for which

$$\lambda(H) + n \neq 0, \quad \forall n \in \mathbf{Z}, \tag{16}$$

for any representative. For a fixed class represented, say, by $\lambda \in \mathcal{H}^*$ the corresponding multiplet consists of an infinite chain of embeddings

$$\cdots \rightarrow \tilde{V}_{-1} \rightarrow \tilde{V}_0 \rightarrow \tilde{V}_1 \rightarrow \cdots \tag{17}$$

where the HWM entering the multiplet are $\tilde{V}_k = V^{\lambda - kN\alpha}$, $k \in \mathbf{Z}$, i.e., they are in 1-1 correspondence with the elements of the class in consideration. Each embedding in (17) is realized by a singular vector $v_s = (X^-)^N \otimes v_0(\tilde{V}_k)$, where $v_0(V)$ denotes the highest weight vector of the HWM V. The factor modules $\tilde{L}_k = \tilde{V}_k/\tilde{V}_{k+1}$ are isomorphic: $\tilde{L}_k \cong \tilde{L}_{k'} \cong \tilde{L}$, $\forall k,k' \in Z$, moreover dim $\tilde{L} = N$ and all states of $\tilde{L}$ are given by $(X^-)^m \otimes v_0$, $m = 0,\ldots,N-1$. Thus the highest weight of an irreducible HWM is determined only up to the equivalence defined above.

3.2.2. The multiplets of the second type are parameterized by a positive integer, say, m such that $m \leq N/2$. Fix such an m and choose an element $\lambda \in \mathcal{H}^*$ such that

$$[\lambda(H) + 1 - m] = 0, \quad m \in \mathbf{N}, \quad m \leq N/2, \tag{18}$$

i.e. $\lambda' = \frac{m-1}{2}\alpha$ is an element of the class of λ. If $m < N/2$ then V^λ is part of an infinite chain of embeddings

$$\cdots \rightarrow V'^{m}_{-1} \rightarrow V^m_0 \rightarrow V'^{m}_0 \rightarrow V^m_1 \rightarrow V'^{m}_1 \rightarrow \cdots \tag{19a}$$

where $V^m_k = V^{\lambda - kN\alpha}$, $k \in \mathbf{Z}$, $V'^{m}_k = V^{\lambda - m\alpha - kN\alpha}$, $k \in \mathbf{Z}$. (Thus the classes which only have elements λ for which (18) holds with $N/2 < m < N$ are represented by the highest weights of V'^{m}_k.) The embeddings $V^m_k \rightarrow V'^{m}_k$ are realized by $v_s = (X^-)^m \otimes v_0(V^m_k)$, while $V'^{m}_k \rightarrow V^m_{k+1}$ are realized by $v_s = (X^-)^{N-m} \otimes v_0(V'^{m}_k)$. The factor modules $L^m_k = V^m_k/V'^{m}_k$ are isomorphic: $L^m_k \cong L^m_{k'} \cong L^m$, $\forall k,k' \in \mathbf{Z}$; also the factor modules $L'^{m}_k = V'^{m}_k/L^m_{k+1}$ are isomorphic: $L'^{m}_k \cong L'^{m}_{k'} \cong L'^m$, $\forall k,k' \in \mathbf{Z}$; moreover dim $L^m = m$, dim $L'^m = N-m$, and all states of L^m (resp. L'^m) are given by $(X^-)^n \otimes v_0$, $n = 0,\ldots,m-1$, (resp. $n = 0,\ldots,N-m-1$). If $N \in 2\mathbf{N}$ and $m = N/2$ then V^λ is part of an infinite chain of embeddings

$$\cdots \rightarrow V^{N/2}_{-1} \rightarrow V^{N/2}_{0} \rightarrow V^{N/2}_{1} \rightarrow \cdots \tag{19b}$$

where $V^{N/2}_{k} = V^{\lambda - kN\alpha/2}$, $k \in \mathbf{Z}$. Everything we said above for L^m_k, L^m is valid here for $m = N/2$.

It is clear that all elements of λ and thus all HWM V^λ over $U_q(\mathcal{G})$ are accounted for in **3.2.1** and **3.2.2**. Thus we have proved:

Proposition 1. *Let* $q^N = 1$, $N \in \mathbf{N}+1$, $\mathcal{G} = sl(2,C)$.

(a) *All* HWM V^λ *over* $U_q(\mathcal{G})$ *belong to multiplets of one of the two types described above.*

(b) *There are exactly* N *inequivalent irreducible* HWM *of* $U_q(sl(2,C))$ *which have dimensions* 1,2,...,N.

The last conclusion was obtained by other methods in [13,14,15]. Note that the non-uniformity in N (denoted there by m) of the results of [13] is due to the fact that their q is a square root of ours.

3.3. Let $\mathcal{G} = sl(3,C)$. Let us denote

$$X^\pm_3 = \pm(q^{1/4}\, X^\pm_1\, X^\pm_2 - q^{-1/4}\, X^\pm_2\, X^\pm_1), \qquad H_3 = H_1 + H_2; \tag{20a}$$

where H_i correspond to the roots α_i; $\alpha_3 = \alpha_1 + \alpha_2 = \rho$; α_1, α_2 are the simple roots with $(\alpha_1,\alpha_2) = -1$; $(\alpha_i,\alpha_i) = 2$, $i = 1,2,3$. We also have (cf. (1),(2)):

$$[H_i, X^\pm_3] = \pm X^\pm_3 \; (= \pm\, \alpha_3(H_i) X^\pm_3), \quad i = 1,2, \tag{20b}$$

$$[X^+_3, X^-_3] = [H_3]. \tag{20c}$$

Note that with this choice, formulae (5), (6) hold also for H_3, $X^\pm_3$. The classification is as follows.

3.3.1. The multiplets of the first type are in 1-1 correspondence with those equivalent classes for which

$$\lambda(H_i) + n \neq 0, \qquad \forall n \in \mathbf{Z}, \qquad \forall i = 1,2,3. \tag{21}$$

for any representative. For a fixed class represented, say, by $\lambda \in \mathcal{H}^*$ the corresponding multiplet consists of the following diagram of embeddings:

$$\begin{array}{ccccccc} & & \vdots & & \vdots & & \\ & & \uparrow & & \uparrow & & \\ \dots & \rightarrow & V_{0,1} & \rightarrow & V_{1,1} & \rightarrow & \dots \\ & & \uparrow & & \uparrow & & \\ \dots & \rightarrow & V_{0,0} & \rightarrow & V_{1,0} & \rightarrow & \dots \\ & & \uparrow & & \uparrow & & \\ & & \vdots & & \vdots & & \end{array} \tag{22}$$

where $V_{k,\ell} = V^{\lambda - kN\alpha_1 - \ell N\alpha_2}$, $k,\ell \in \mathbf{Z}$, are again parameterized by the elements equivalent to λ. These equivalence statements will be omitted below. Each embedding in (22) is realized by a singular vector

$$v_s = (X_i^-)^N \otimes v_0, \text{ for } i = 1,2, \text{ resp.,}$$

when the arrow depicting the embedding is horizontal or vertical, resp. Because of the symmetry it is clear that the factor modules $L_{k,\ell} = V_{k,\ell}/I_{k,\ell}$, where $I_{k,\ell}$ is the maximal submodule of $V_{k,\ell}$, are isomorphic:

$$L_{k,\ell} \cong L_{k',\ell'} \cong L^1, \forall k,k',\ell,\ell' \in \mathbf{Z}.$$

For lack of space we omit the proof that $\dim L^1 = N^3$.

In (22) and in all diagrams below we do not depict any embeddings outside the quadrangle $(V_{0,0}, V_{1,0}, V_{1,1}, V_{0,1})$ except the adjacent ones shown in (22).

3.3.2. The multiplets of the second type are parameterized by a positive integer, say, m_1 such that $m_1 \leq N/2$. Fix such an m_1 and choose an element $\lambda \in \mathcal{H}^*$ such that

$$[\lambda(H_1) + 1 - m_1] = 0, \quad \lambda(H_i) + n \neq 0, \quad i = 2,3, \quad \forall n \in \mathbf{Z}. \tag{23}$$

If $m_1 < N/2$ then V^λ is part of the following multiplet:

$$
\begin{array}{ccccccccc}
 & & \vdots & & & & \vdots & & \\
 & & \uparrow & & & & \uparrow & & \\
\cdots & \to & V_{0,1} & \to & V^1_{0,1} & \to & V_{1,1} & \to & \cdots \\
 & & \uparrow & & \uparrow & & \uparrow & & \\
\cdots & \to & V_{0,0} & \to & V^1_{0,0} & \to & V_{1,0} & \to & \cdots \\
 & & \uparrow & & & & \uparrow & & \\
 & & \vdots & & & & \vdots & &
\end{array}
\qquad (24)
$$

where $V_{k,\ell}$ are as above and $V^1_{k,\ell} = V^{\lambda - m_1\alpha_1 - kN\alpha_1 - \ell N\alpha_2}$, $k,\ell \in \mathbf{Z}$. The embeddings $V_{k,\ell} \to V^1_{k,\ell}$ are realized by $v_s = (X^-_1)^m \otimes v_0(V_{k,\ell})$, while $V^1_{k,\ell} \to V_{k+1,\ell}$ are realized by $v_s = (X^-_1)^{N-m} \otimes v_0(V^1_{k,\ell})$. The factor modules $L_{k,\ell} = V_{k,\ell}/I_{k,\ell}$ are isomorphic: $L_{k,\ell} \cong L_{k'\ell'} \cong L^2_m$, $\forall k,\ell,k',\ell' \in \mathbf{Z}$. For lack of space we omit the proof that $\dim L^2_{m_1} = m_1N^2$, $\dim L'^2_{m_1} = (N - m_1)N^2$. If $N \in 2N$ and $m_1 = N/2$ then everything said above for $L^2_{m_1}$ is valid here for $m_1 = N/2$. Thus there are N-1 irreducible HWM with highest weights satisfying (23).

We do not consider separately the subcase obtained from this by exchanging the indices 1 and 2. The corresponding representations which are conjugate to $L^2_{m_1}$, $L'^2_{m_1}$, resp., under the exchange $\alpha_1 \leftrightarrow \alpha_2$ will be denoted by $\tilde{L}^2_{m_2}$, $\tilde{L}'^2_{m_2}$, resp., with $\dim \tilde{L}^2_{m_2} = m_2N^2$, $\dim \tilde{L}'^2_{m_2} = (N - m_2)N^2$, resp.

3.3.3. The multiplets of the third type are parameterized by a positive integer, say, m_3 such that $m_3 \leq N/2$. Fix such an m_3 and choose an element $\lambda \in \mathcal{H}^*$ such that

$$[\lambda(H_3) + 2 - m_3] = 0, \quad \lambda(H_i) + n \neq 0, \quad i = 1,2, \quad \forall n \in \mathbf{Z}. \qquad (25)$$

The HWM V^λ is part of the following multiplet:

$$\begin{array}{ccccccccc}
 & & \vdots & & & & \vdots & & \\
\cdots & \rightarrow & V_{0,1} & \rightarrow & & \rightarrow & V_{1,1} & \rightarrow & \cdots \\
 & & & & & \nearrow & & & \\
 & & \uparrow & & V^3_{0,0} & & \uparrow & & \\
 & & & \nearrow & & & & & \\
\cdots & \rightarrow & V_{0,0} & \rightarrow & & \rightarrow & V_{1,0} & \rightarrow & \cdots \\
 & & \uparrow & & & & \uparrow & & \\
 & & \vdots & & & & \vdots & &
\end{array} \tag{26}$$

where $V_{k,\ell}$ are as above and $V^3_{k,\ell} = V^{\lambda - m_3\alpha_3 - kN\alpha_1 - \ell N\alpha_2}$, $k,\ell \in \mathbf{Z}$. The embeddings $V_{k,\ell} \rightarrow V^3_{k,\ell}$ and $V^3_{k,\ell} \rightarrow V_{k+1,\ell+1}$ are realized by the following singular vector:

$$v^p_s = \sum_{j=0}^{p} a_j\, (X^-_1)^{p-j}\, (X^-_2)^p\, (X^-_1)^j \otimes v_0, \tag{27a}$$

$$a_j = (-1)^j a \frac{[\lambda(H_1) + 1]}{[\lambda(H_1) + 1 - j]} \binom{p}{j}_q, \qquad j = 0,\dots,p, \quad a \neq 0, \tag{27b}$$

(or by the same expression with the indices 1 and 2 interchanged), for $p = m_3$, $p = N-m_3$, respectively. Formula (27) can be checked directly. It is valid for any $p \in \mathbf{N}$, if (25) holds (with m_3 replaced by p), however, if $p \geq N$, and $p = kN+t$, $k \in \mathbf{N}$, $t \in \mathbf{Z}_+$, $t < N$ it reduces to: $v^p_s = (X^-_1)^{kN} (X^-_2)^{kN} v^t_s$. It holds in general in all cases when α_3 is the sum of two roots α_1, α_2 with equal length. For $q \rightarrow 1$ it goes to the correct formula in the same situation [17] [28] (cf. (8.40), (8.41)) and arbitrary p. Analogously to the previous case there are N-1 inequivalent irreducible HWM with highest weights satisfying (25), namely the factor modules

$$L^3_{m_3} \cong V_{k,\ell}/I_{k,\ell} \ \forall k,\ell \in Z,\ m_3 \leq N/2;\ L'^3_{m_3} \cong V^3_{k,\ell}/I^3_{k,\ell} \ \ \forall k,\ell \in Z,\ m_3 < N/2.$$

We omit the proof that $\dim L^3_{m_3} = m_3N^2$, $\dim L'^3_{m_3} = (N-m_3)N^2$.

3.3.4. The multiplets of the fourth type are parameterized by two positive integers, say, m_1, m_2 such that $m_1 + m_2 < N$. Fix such m_1, m_2 and choose an element $\lambda \in \mathcal{H}^*$ such that

$$[\lambda(H_i) + 1 - m_i] = 0, \quad i = 1,2, \ \Rightarrow \ [\lambda(H_3) + 2 - m_1 - m_2] = 0. \tag{28}$$

The HWM V^λ is part of the following multiplet:

$$\begin{array}{cccccccccc}
 & & \vdots & & & & & \vdots & & \\
 & & \uparrow & & & & & \uparrow & & \\
\cdots & \rightarrow & V_{0,1} & \rightarrow & V^1_{0,1} & & \rightarrow & V_{1,1} & \rightarrow & \cdots \\
 & & & & \uparrow & & \nearrow & & & \\
 & & \uparrow & & V^{12}_{0,0} & \rightarrow & V^3_{0,0} & \uparrow & & \\
 & & & \nearrow & \uparrow & & \uparrow & & & \\
 & & V^2_{0,0} & \rightarrow & & \rightarrow & V^{21}_{0,0} & \rightarrow V^2_{1,0} & & \\
 & & & & \uparrow & \nearrow & & & & \\
 & & \uparrow & & & & & \uparrow & & \\
\cdots & \rightarrow & V_{0,0} & \rightarrow & V^1_{0,0} & & \rightarrow & V_{1,0} & \rightarrow & \cdots \\
 & & \uparrow & & & & & \uparrow & & \\
 & & \vdots & & & & & \vdots & &
\end{array} \tag{29}$$

where $V_{k,\ell}$ are as before and

$$V^i_{k,\ell} = V^{\lambda - m_i\alpha_i - kN\alpha_1 - \ell N\alpha_2}, \quad i = 1,2,3, \quad m_3 = m_1 + m_2, \quad k,\ell \in Z, \tag{30a}$$

$$V^{ij}_{k,\ell} = V^{\lambda - m_i\alpha_i - m_3\alpha_j - kN\alpha_1 - \ell N\alpha_2}, \quad (ij) = (12),(21), \quad k,\ell \in \mathbf{Z}. \tag{30b}$$

We summarize the structure of the above multiplets as follows.

3.3.4.1. The embeddings $V_{k,\ell} \to V^1_{k,\ell}$, $V^1_{k,\ell} \to V_{k+1,\ell}$, $V^2_{k,\ell} \to V^{21}_{k,\ell}$, $V^{21}_{k,\ell} \to V^2_{k+1,\ell}$, $V^{12}_{k,\ell} \to V^3_{k,\ell}$, $V^3_{k,\ell} \to V^{12}_{k+1,\ell}$, $V_{k,\ell+1} \to V^1_{k,\ell+1}$, $V^1_{k,\ell+1} \to V^1_{k+1,\ell+1}$, resp., are realized by singular vectors $(X^-_1)^p \otimes v_0$ with $p = m_1$, $N-m_1$, m_3, $N-m_3$, m_2, $N-m_2$, m_1, $N-m_1$, resp. The embeddings $V_{k,\ell} \to V^2_{k,\ell}$, $V^2_{k,\ell} \to V_{k,\ell+1}$, $V^1_{k,\ell} \to V^{12}_{k,\ell}$, $V^{12}_{k,\ell} \to V^1_{k,\ell+1}$, $V^{21}_{k,\ell} \to V^3_{k,\ell}$, $V^3_{k,\ell} \to V^{21}_{k,\ell+1}$, $V_{k+1,\ell} \to V^2_{k+1,\ell}$, $V^2_{k+1,\ell} \to V_{k+1,\ell+1}$, resp., are realized by singular vectors $(X^-_2)^p \otimes v_0$ with

$$p = m_2,\ N-m_2,\ m_3,\ N-m_3,\ m_1,\ N-m_1,\ m_2,\ N-m_2, \text{ resp.}$$

The embeddings $V^1_{k,\ell} \to V^{21}_{k,\ell}$, $V^2_{k,\ell} \to V^{12}_{k,\ell}$, $V^3_{k,\ell} \to V_{k+1,\ell+1}$, resp., are realized by singular vectors given by formula (27) with λ replaced by $\lambda - m_1\alpha_1$, $\lambda - m_2\alpha_2$, $\lambda - m_3\alpha_3$, resp., and $p = m_2$, $p = m_1$, $p = N-m_3$, resp.

3.3.4.2. Note that the six HWM $V_{k,\ell}$, $V^i_{k,\ell}$, $V^{ij}_{k,\ell}$ for fixed k,ℓ form the basic sl(3,C) multiplet in the case $q = 1$, (cf. [18, formula (38)]). The sextet diagram consisting of these six HWM is commutative which one checks also here using formula (27). Let us say that the tip of this sextet is at $V_{k,\ell}$. This sextet shares one side with six sextets of the same type and orientation and for the same k,ℓ their tips are at $V^{21}_{k-1,\ell-1}$, $V^{12}_{k,\ell-1}$, $V^{21}_{k,\ell}$, $V^{12}_{k,\ell}$, $V^{21}_{k-1,\ell}$, $V^{12}_{k-1,\ell-1}$. The role of (m_1, m_2) in these sextets is played by $(m_2, N-m_3)$ for $V^{12}_{*,*}$ and by $(N - m_3, m_1)$ for $V^{21}_{*,*}$.

(As a side remark we note that there are other sextets of HWM, namely: $V^3_{k-1,\ell-1}$, $V^{12}_{k,\ell-1}$, $V^2_{k+1,\ell}$, $V_{k+1,\ell+1}$, $V^1_{k,\ell+1}$, $V^{21}_{k-1,\ell}$, for fixed k,ℓ and containing the sextet $V_{k,\ell}$, $V^i_{k,\ell}$, $V^{ij}_{k,\ell}$. However, these bigger sexters are more complicated than the smaller sl(3,C)-like sextets.)

3.3.4.3. Thus the structure of the representations $V_{k,\ell}$, $V^{12}_{k,\ell}$, $V^{21}_{k,\ell}$, is exactly the same; moreover the range of their parameters is the same. The same holds for the representations $V^{i}_{k,\ell}$, $i = 1,2,3$. These are situated in the sextets at the site opposite to what we called the tip. The values $(\lambda(H_1), \lambda(H_2))$, i.e., the analogues of (m_1,m_2), are $(N\text{-}m_1, m_3)$, $(m_3, N\text{-}m_2)$, $(N\text{-}m_2, N\text{-}m_1)$, resp., for $i = 1,2,3$, resp., and they cover the same range. Moreover, this shows that the requirement $m_1+m_2 < N$ is not a restriction. Indeed, the HWM $V^{i}_{k,\ell}$ for one value of i exhaust all such cases.

3.3.4.4. From the above it is easy to see that there are the following inequivalent irreducible HWM with highest weights satisfying (28), namely the factor modules $L^{4}_{m_1m_2} \cong V_{k,\ell}/I_{k,\ell}$, $L'^{4}_{m_1m_2} \cong V^{3}_{k,\ell}/I^{3}_{k,\ell}$. For the lack of space we omit the proof that $\dim L^{4}_{m_1m_2} = m_1m_2(m_1+m_2)/2$. In particular, L^{4}_{11} is the trivial 1-dimensional representation. Note that when $q = 1$ the last formula, however, without any restriction on $m_1, m_2 \in \mathbf{N}$, gives the dimensions of *all* finite dimensional irreducible HWM of $sl(3,\mathbf{C})$ (or equivalently of all unitary irreducible representations of $su(3)$ or the group $SU(3)$).

The multiplets of the next type can be viewed as "analytic" continuation in m_i for $m_1+m_2 = N$.

3.3.5. The multiplets of the fifth type are parameterized by a positive integer, say, m_1 such that $m_1 \leq N/2$. Fix such m_1 and choose an element $\lambda \in \mathcal{H}^*$ such that

$$[\lambda(H_i) + 1 - m_i] = 0, \quad m_2 = N\text{-}m_1. \tag{31}$$

The HWM V^{λ} is part of a multiplet containing the following HWM: $V_{k,\ell}$ and $V^{i}_{k,\ell}$, $i = 1,2$ given by the same formulae as in the previous case with $m_2 = N\text{-}m_1$ and $m_3 = N$. It can be depicted using (29) and distorting it so that $V^{3}_{k,\ell}$ will coincide with $V_{k+1,\ell+1}$, $V^{12}_{k,\ell}$ with $V^{2}_{k+1,\ell}$, $V^{21}_{k,\ell}$ with $V^{1}_{k,\ell+1}$. Thus the sextets with $V^{12}_{k,\ell}$, $V^{21}_{k,\ell}$ at the tips deteriorate into commutative triangles and the latter representations do not have the structure of $V_{k,\ell}$. The singular vectors depicting the embeddings are as in the previous case, however, taking into account the coincidences. It is easy to see that there are the following inequivalent irreducible

HWM with highest weights satisfying (31), namely the factor modules $L^5_{m_1} \equiv V_{k,\ell}/I_{k,\ell}$, $L^{51}_{m_1} \equiv V^1_{k,\ell}/I^1_{k,\ell}$, $L^{52}_{m_1} \equiv V^2_{k,\ell}/I^2_{k,\ell}$. Note that $L^{51}_{m_1}$, $L^{52}_{m_2}$ are conjugate to each other under the exchange $\alpha_1 \longleftrightarrow \alpha_2$. For lack of space we omit the proof that $\dim L^5_{m_1} = m_1(N-m_1)N/2$.

Summarizing everything in Subsection 3.3 we have proved:

Proposition 2. *Let* $q^N = 1$, $N \in \mathbb{N}+1$, $\mathcal{G} = sl(3,\mathbb{C})$.

(a) *All* HWM V^λ *over* $U_q(\mathcal{G})$ *belong to multiplets of one of the five types described above.*

(b) *The list of inequivalent irreducible* HWM *of* $U_q(sl(3,\mathbb{C}))$ *consists of the factor-modules* L^1 *from* 3.3.1; $L^2_{m_1}$ *for* $m_1 \le N/2$, $L'^2_{m_1}$ *for* $m_1 < N/2$, $\tilde{L}^2_{m_2}$ *for* $m_1 \le N/2$, $\tilde{L}'^2_{m_1}$ *for* $m_2 < N/2$, *from* 3.3.2; $L^3_{m_3}$ *for* $m_3 \le N/2$, $L'^3_{m_3}$ *for* $m_3 < N/2$, *from* 3.3.3; $L^4_{m_1 m_2}$, $L'^4_{m_1 m_2}$, *from* 3.3.4; $L^5_{m_1}$, $L^{51}_{m_1}$, *for* $m_1 \le N/2$, $L^{52}_{m_1}$ *for* $m_1 < N/2$, *from* 3.3.5.

Acknowledgements

The author would like to thank Professor Abdus Salam, the International Atomic Energy Agency and UNESCO for hospitality at the International Centre for Theoretical Physics, Trieste. He would like to thank Dr V B Petkova for stimulating discussions. This work was partially supported by the Ministry of Science, Culture and Education of Bulgaria under Grants 3 and 403.

References

1. L D Fadeev, Integrable models in 1+1 dimensional quantum field theory, in *Les Houches Lectures 1982* (Elsevier, Amsterdam, 1984).

2. P P Kulish & E K Sklyanin, *Lecture Notes in Physics* **151** (1982), 61-119.

3. P P Kulish & N Yu, *Reshetikhin, Zap. Nauch. Semin. LOMI* **101** (1981), 101-110 (in Russian); English translation: *J. Soviet Math.* **23** (1983), 2435-2441.

4. E K Sklyanin, (a) *Funkts. Anal. Prilozh.* **16** (1982), 27-34, **17** (1983), 34-48 (in Russian); English translation: *Functional Anal. Appl.* **16** (1982), 263-270, **17** (1983), 274-288; (b) *Uspekhi Mat. Nauk* **40** (1985), 214 (in Russian).

5. V G Drinfeld, *Dokl. Akad. Nauk SSSR* **283** (1985), 1060-1064 (in Russian); English translation: *Soviet Math. Dokl.* **32** (1985), 254-258; Proceedings ICM (MRSI, Berkeley, 1986).

6. N Yu Reshetikhin, Quantized universal enveloping algebras, the Yang-Baxter equation and invariants of links I.II., *LOMI Leningrad preprints E-4-87, E-17-87* (1987).

7. A N Kirillov & N Yu Reshetikhin, Representations of the algebra $U_q(sl(2))$, q-orthogonal polynomials and invariants of links, *LOMI Leningrad preprint E-9-88* (1988).

8. M Jimbo, *Lett. Math. Phys.* **10** (1985), 63-69; *Lett. Math. Phys.* **11** (1986), 247-252.

9. M Rosso, *C. R. Acad. Sci. Paris Sér. I Math.* **305** (1987), 587-590.

10. S L Woronowicz, *Comm. Math. Phys.***111** (1987), 613-665.

11. M Rosso, *Comm. Math. Phys.* **117** (1987), 581-593.

12. G Lusztig, *Adv. Math.* **70** (1988), 237-249.

13. P Roche & D Arnaudon, *Irreducible representations of the quantum analogue of SU(2)* (Ecole Polytechnique Palaiseau preprint 1988).

14. L Alvarez-Gaumé, C Gomez & G Sierra, *Phys. Lett.* **220B** (1989), 142-152.

15. V Pasquier & H Saleur, Saclay preprint SPhT/89-031 (1989).

16. L Alvarez-Gaumé, C Gomez & G Sierra, preprint CERN-TH.5129/88 (1988).

17. V K Dobrev, *J. Math. Phys.* **26** (1985), 235-251 and ICTP Trieste preprint IC/83/36 (1983); Talk at the I National Congress of Bulgarian Physicists, Sofia (1983), INRNE Sofia preprint (1983) and *Lett. Math. Phys.* **9** (1985), 205-211.

18. V K Dobrev, Talk at the Conference on Algebraic Geometry and Integrable Systems, Oberwolfach (July 1984) and ICTP Trieste preprint IC/85/9 (1985).

19. V K Dobrev, in *Proceedings of the XIII International Conference on Differential-Geometric Methods in Theoretical Physics, Shumen (1984)* (Eds H D Doebner and T D Palev, World Sci., Singapore, 1986), 348-370 and ICTP Trieste preprint IC/85/13 (1985).

20. V K Dobrev & V B Petkova, *Lett. Math. Phys.* **9** (1985), 287-298.

21. V K Dobrev, *Lett. Math. Phys.* **11** (1986), 225-234.

22. E Abe, *Hopf Algebras* (Cambridge Tracts in Math., **74**, Cambridge Univ. Press, 1980).

23. J Dixmier, *Enveloping Algebras* (North Holland, New York, 1977).

24. I N Bernstein, I M Gel'fand & S I Gel'fand, *Funkts. Amal. Prilozh.* **5** (1971), 1-9; English translation: *Funct. Anal. Appl.* **5** (1971), 1-8.

25. V G Kac & D Kazhdan, *Adv. Math.* **34** (1979), 97-108.

26. F G Malikov, B L Feigin & D B Fuchs, *Funkts. Anal. Prilozh.* **20** (1986), 25-37; English translation: *Funct. Anal. Appl.* **20** (1986), 103-113.

27. V G Kac, *Infinite-dimensional Lie algebras. An introduction* (Progr. Math. **44** Birkhäuser, Boston, 1983).

28. V K Dobrev, *Reports Math. Phys.* **25** (1988), 159-181 and ICTP Trieste internal report IC/86/393 (1986).

SOLUTIONS OF CERTAIN SETS OF EQUATIONS OVER GROUPS

M EDJVET

University of Nottingham, University Park, Nottingham NG7 2RD

Introduction

Let G be a group, F the free group generated by t and let r(t) be an element of the free product G*F. The equation r(t) = 1 is said to have a solution over G if it has a solution in some group that contains G. This is equivalent to saying that the natural map $G \to \frac{G*F}{N}$, where N denotes the normal closure in G*F of r(t) is injective. The Kervaire-Laudenbach (KL) conjecture asserts that if the exponent sum of t in r(t) is non-zero then this natural map is indeed a monomorphism.

The KL conjecture has received much attention but still remains unsettled. For example in recent years S M Gersten, J Howie and J R Stallings have between them produced nine conjectures any of which would have implied the truth of the KL conjecture only for Gersten to produce a counter-example to all nine! (We refer the reader to [5] and the references cited there.) When Gersten's counter-example is translated into our present situation the element r(t) is at^2dt^{-1} ($a,d \in G$), in which case r(t) = 1 has a solution over G, [6]. What does all this mean? Gersten suggests in [5] that we should be looking for a counter-example. The problem with this is that it is difficult to know where to look. So for the moment we content ourselves with verifying the KL conjecture for specific forms of the word r(t).

Our interest in this problem started with some joint work with Howie in which we proved the following result.

Theorem A. *Let* G *be a group, and* $a,b,c,d \in G$. *Then the equation*

$$atbtctdt^{-1} = 1$$

has a solution over G.

The length of the equation r(t) = 1 is the sum of the absolute values of the exponents of t in r(t). It follows from Theorem A together with earlier results by Howie [6] and Levin [7] that the KL conjecture is true for when the length of r(t) is at most 4.

Part of the proof of Theorem A deals with when r(t) is at^3dt^{-1}. This has led the present author on to a study of the problem for when r(t) is the word at^ndt^{-m} ($n,m \in \mathbf{Z}$, $n \neq m$). In this paper we shall prove the following result.

Theorem B. *Let* G *be a group and* $a,d \in G$. *Then the equation* at^ndt^{-1} ($n \neq 1$) *has a solution over* G.

Observe that

$$< G,t \mid at^{np}dt^{-p} > \; = \; < G,t,s \mid at^{np}dt^{-p}, st^{-p} >$$

$$= \; < G,t,s \mid as^nds^{-1}, st^{-p} > \; = \; < G,s \mid as^nds^{-1} > \underset{s=t^p}{*} < t \mid t^{\lambda} >,$$

where $\lambda = 0$, pq (respectively) if s has infinite order, order q (respectively) in $< G,s \mid as^nds^{-1} >$, whence we obtain the following consequence of Theorem B.

Corollary. *The equation* $at^{np}dt^{-p}$ ($np \neq p$) *has a solution over* G.

We announce the following two results, the proofs of which shall appear in [3].

Theorem C. *Let* G *be a group, and* a,d *be elements of order at least* 3 *in* G. *Then the equation* at^ndt^{-m} ($n \neq m$) *has a solution over* G.

Theorem D. *Let* G *be a group and let* a,d *be elements of* G *one of which has order* 2 *and the other not having order* 3 *in* G. *Then the equation* at^ndt^{-m} ($n \neq m$) *has a solution over* G.

Observe that if the equation r(t) = 1 has a solution over the subgroup $< a,d >$ of G generated by a and d in some overgroup H, say, then it has a solution over G in the group $G \underset{<a,d>}{*} H$, so it may be assumed that $G = < a,d >$. It follows from all this that the KL conjecture is true for when r(t) is at^ndt^{-m} apart from the case when G is a quotient of the modular group. The Gerstenhaber-Rothaus theorem [8], says that if G is locally residually finite then the KL conjecture is true, so in fact the above case is settled apart from when G is a non-residually finite quotient of the modular group (which, of course, has such quotients since, for example, it is SQ-universal).

In Section 1 we introduce some notation and the main technique to be used, that is, so-called relative diagrams. This section has been kept rather brief and it may help the reader to refer to Section 1 of [6] for further details. In Section 2 we prove Theorem B.

1. Preliminaries

As mentioned in the introduction it can henceforth be assumed that $G = < a,d >$ and that G is not a residually finite group. It can be assumed further without any loss, after replacing r(t) and t by their inverses and cyclically permuting, if necessary, that the order of a in G is less than or equal to the order of d in G, and, by Levin's theorem [7], that $n > 0$. Since the case $n = 2$ was done in [6] and $n = 3$ in [4] we let $n \geq 4$ in Theorem B.

For the purposes of the proof, we assume that the element at^ndt^{-1} takes the form

$$atx_2t \dots x_ntdt^{-1} \quad (n \geq 4)$$

where $x_j = 1$ $(2 \leq j \leq n)$.

Let D be a *relative diagram* [6] for the equation $at^ndt^{-1} = 1$. Then D is a tesselation of the 2-sphere and has a distinguished vertex V_0. The vertices of D other than V_0 are called *interior* vertices. The faces of D are all (n+1)-sided with edge-orientations and *corner-labels* as illustrated (up to cyclic permutation and inversion) in Fig. 1.1. A face will be called interior if all its vertices are interior.

The *label* of a vertex V of D is the list of corner labels at V read in a counter-clockwise direction from any starting point to make a word in

$$\{a, x_j \ (2 \leq j \leq n), d\}.$$

The label of each vertex $V \neq V_0$ is a relator in the group G. The label of V_0 is a non-trivial element of G. As in [4] the vertex labels correspond to closed paths in the associated "labelled star graph" Γ of Fig. 1.2. This fact helps when listing possible vertex labels.

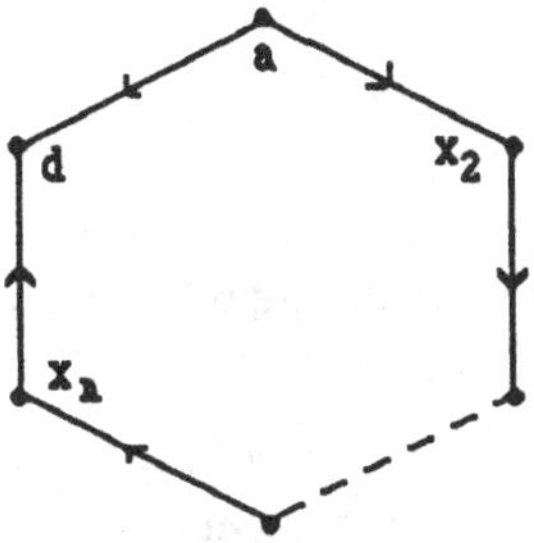

Fig. 1.1

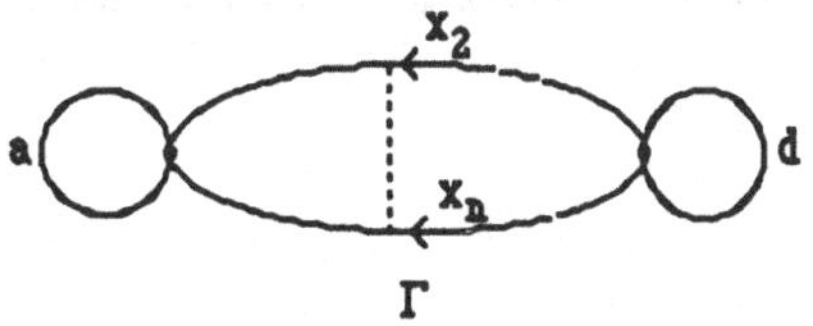

Fig. 1.2

Our methods involve a detailed study of faces and vertices of various relative diagrams. When we gave lists of these in what follows we will do so *up to cyclic permutation and inversion*. With this understood and, for ease of presentation, we omit mentioning this phrase for the remainder of the text.

Notation. If $g \in G$ then $|g|$ shall denote the order of g in G. In Section 2 we use ω to denote a reduced word (possibly empty) in the x_j $(2 \leq j \leq n)$. We emphasize here that ω does not necessarily denote the same reduced word each time it occurs, even within the same label. For example, in the label $(\omega d)^3$, the three occurences of ω could be distinct. A vertex of a relative diagram whose label does not involve either a or d will be called an x-*vertex* ; a vertex whose label involves a only will be called a *source*; and a vertex whose label involves d only will be called a *sink*. By the term *lane* we shall mean a path in a relative diagram maximal with respect to the property that every intermediate vertex is interior and of degree 2. (Note that it is possible for D to have vertices of degree 2 labelled by $a^{\pm 2}$ or $d^{\pm 2}$ and these are given by Fig. 1.3; observe that the adjacent vertices in each case are of degree at least 3 and involve a or d (or both).)

Fig. 1.3

Finally in this section, we present two results, one concerning relative diagrams and the other group-theoretic, which will be needed later.

Lemma 1.1. *Let* D *represent a counter-example to Theorem B. Suppose that* UV *and* VU *in Fig. 1.4(i) are lanes in* D *of equal length (path-length metric) all of whose intermediate vertices are* x-*vertices. Suppose further that the sublabel* z_3 *of* V *is a word of even length in the* x_j. *Then we can obtain a new relative diagram* D' *representing a counter-example to Theorem B which has fewer faces than* D.

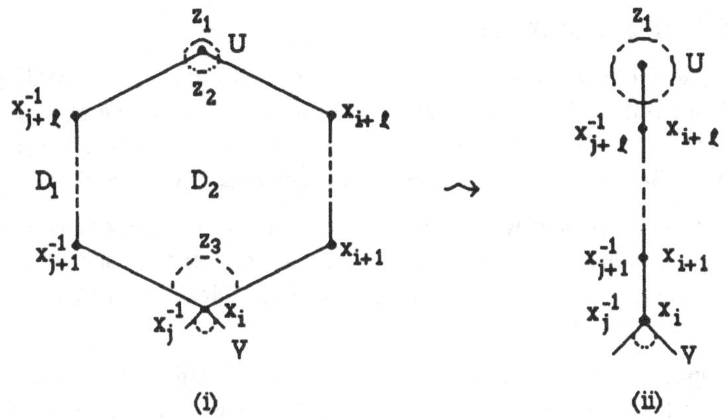

Fig. 1.4

Proof. Observe that since z_3 has even length and the intermediate vertices of UV and VU are x-vertices, the vertex sublabel z_2 (and so z_1) corresponds to a closed path in Γ.

Suppose firstly that $U \neq V_0$, the distinguished vertex of D. If $z_1 = 1$ in G then what we do is simply delete from D that subdiagram D_1 or D_2 *not* containing V_0 and 'sew up' the boundary UVU to obtain a counter-example with fewer faces. By the term 'sew up' we mean identify the edges and vertices of the lanes UV and VU in the obvious way (see Fig. 1.4(ii)). If $z_1 \neq 1$ in G then delete that subdiagram containing V_0 and 'sew up' to obtain a relative diagram with distinguished vertex U', the vertex obtained from U, representing a counter-example with fewer faces.

Suppose now that $U = V_0$. If $z_1 = 1$ in G then $z_2 \neq 1$ in G, and so we can delete D_1 and 'sew up' as before. If $z_2 = 1$ in G then we do the same only this time D_2 is deleted.

Lemma 1.2 *Let* G *be generated by* a *and* d. *If any of the following conditions hold then* G *is abelian-by-finite (in particular, every quotient of* G *is residually finite)*:

(i) $a^2 = (ad)^3 = 1$ *and* $|d| \leq 6$;

(ii) $a^2 = d^3 = 1$ *and* $ad^{\delta_1}...ad^{\delta_\ell} = 1$ *in* G, *where* $\delta_j = \pm 1$ *for* $1 \leq j \leq \ell$ *and* $1 \leq \ell \leq 6$.

We omit the proof of Lemma 1.2 (part (ii) of which can be deduced, for example, from results in [2]).

2. Proof of Theorem B

Assume by way of contradiction that the theorem is wrong. Then according to Lemma 1 of [6] there is a relative diagram D for $at^n dt^{-1} = 1$ which represents a counter-example. Let N_1 denote the total number of sources and sinks in any given relative diagram, and let N_2 denote the total number of vertices of degree 2.

We make the assumption that D contains the least number of faces of any relative diagram representing a counter-example. Assume further that, subject to this minimum number of faces, the number N_1 is maximal. Finally assume, subject to these two constraints, that N_2 is maximal.

The assumption minimum number of faces forces the vertex labels of D to be cyclically reduced words, and, furthermore, it implies that on deleting any sublabel equal to 1 in G from any vertex label, the word that remains is cyclically reduced. (We refer the reader to [6] for the proofs of these assertions.)

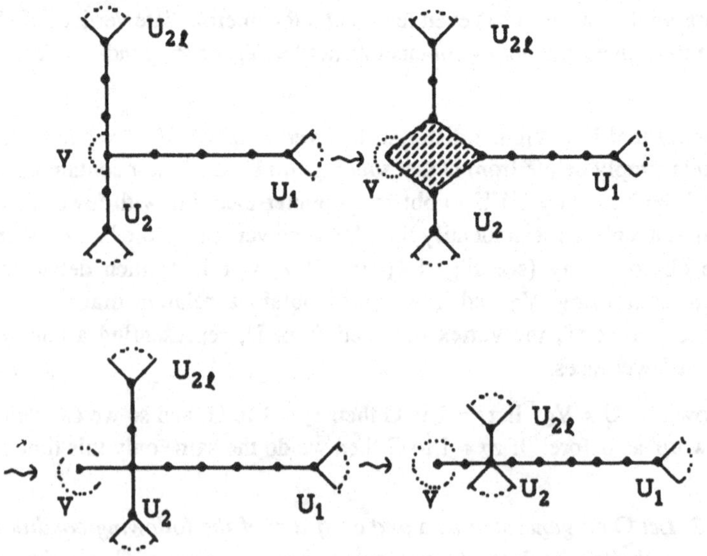

Fig. 2.1

Lemma 2.1. D *contains no vertices of degree* > 2 *whose label involves only the* x_j $(2 \le j \le n)$.

Proof. For suppose V were such a vertex. Then deg V = 2ℓ where $\ell \ge 2$. Let U_i $(1 \le i \le 2\ell)$ be the 2ℓ vertices of degree at least 3 that are the endpoints (other than V) of the 2ℓ distinct lanes incident to V. (Observe from Fig. 1.3 that since V is an x-vertex, the intermediate vertices of the lane VU_i are x-vertices, and, furthermore, U_i is neither a source nor a sink $(1 \le i \le \ell)$.) We show by induction on ℓ that we can obtain a new diagram which represents a counter-example which has the same number of faces as D, but a smaller total number of vertices of degree 2, a contradiction. In fact the new diagram is obtained by applying to D a sequence of diamond moves. (For a discussion of *diamond moves* see [1] and [4].) Suppose that the result holds for $2 \le \ell \le k$, and let deg V = 2k+2. The sequence of diamond moves indicated by the example in Fig. 2.1 shows the way to reduce to case deg V = 2k. (Observe that the total number of vertices of degree 2 is at first decreased but restored by the final move in the sequence.) If $\ell = 2$ we obtain a contradiction by a similar sequence of diamond moves.

The above argument breaks down if, for example, in Fig. 2.1(ii) U_2 and $U_{2\ell}$ are the same vertex, and the length of VU_2 equals that of $VU_{2\ell}$. But by Lemma 1.1 we can then obtain a new relative diagram representing a counter-example which has fewer faces than D, a contradiction.

2.1. $3 \le |a| \le |d|$

For a vertex V of D we shall be interested in the number of edges incident to both V and to interior vertices of degree 2 that can be arranged consecutively around V. This will correspond to some sublabel, s(V) say, of the label of V. Now although s(V) can in general have subwords on the x_j of arbitrary length it is easy to see that since $2 < |a|, |d|$, s(V) can involve at most one a and at most one d. The case when both a and d are involved is illustrated in Fig. 2.2.

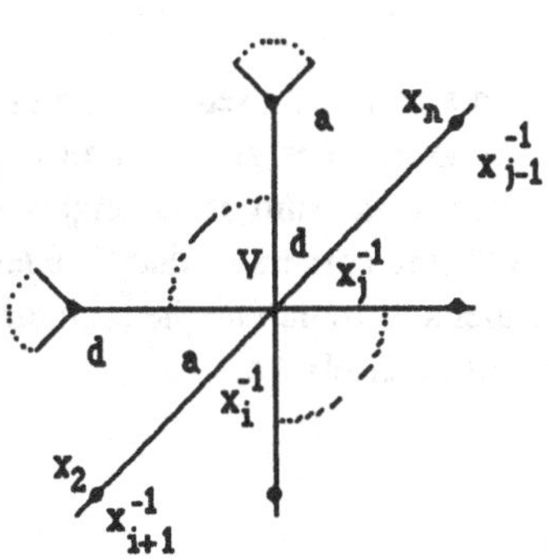

Fig. 2.2

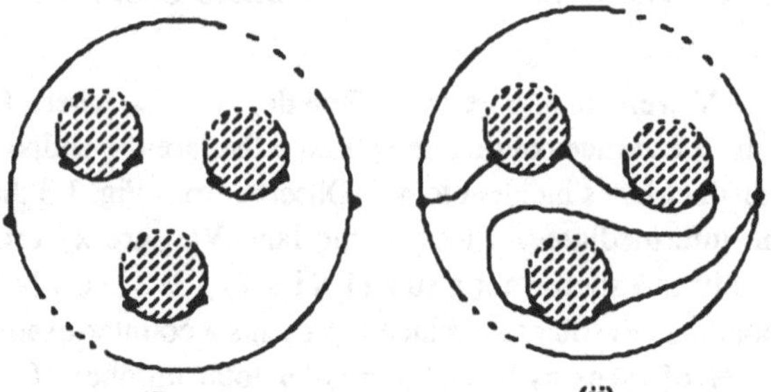

(i) (ii)

Fig. 2.3

Obtain a new diagram D' by removing from D all interior vertices of degree 2 and edges incident to them. It follows from Lemma 2.1 that D' contains no isolated vertices. It may however contain isolated subdiagrams, see, for example, Fig. 2.3(i). What we do is simply reintroduce the least amount of lanes that had been removed as is required to reconnect the diagram. Finally obtain the diagram $\tilde{D}$ by replacing in D' each of the reintroduced lanes by a line segment. This way we get rid of any vertices of degree 2 these lanes may have contained. (See Fig. 2.3(ii).)

The basic list of corner labels

$$a^{\pm 1}, x_j^{\pm 1} \ (2 \leq j \leq n), d^{\pm 1}$$

of D have been supplemented in forming $\tilde{D}$. From the above paragraphs we see that these new basic sublabels are the subwords of $(a\omega d)^{\pm 1}$. Of course there will be infinitely many possibilities for the word ω; we are, however, only interested in the fact that ω does not involve a or d.

We shall use the same curvature arguments as those used in [4] and [6]. The corner of any face of a given relative diagram D is assigned an angle. The curvature of a vertex V is defined to be 2π less the sum of the angles at V. The curvature of a k-gonal face is the sum of all the corners of that face less $(k-2)\pi$. With these definitions, it follows from Euler's formula that the total curvature of D, summed over all the vertices and faces, is precisely 4π.

Let each corner at a vertex of degree ℓ have angle $2\pi/\ell$. Then the vertices of D have zero curvature and we need only consider faces. If P is an m-gon of D and the degrees of the vertices of P are m_i ($1 \le i \le m$) then the curvature contributed by P to the total curvature of D is

$$(2-m)\pi + 2\pi \sum_{i=1}^{m} \frac{1}{m_i}. \qquad (*)$$

Lemma 2.2. *Each interior vertex of* $\tilde{D}$ *has degree at least* 3 *and each face has degree at least* 4.

Proof. We already know from Lemma 2.1 that deg $V \ge 1$ for each vertex V of $\tilde{D}$. If deg V = 1 or 2 then an easy check shows that the relator obtained from the label of V must be one of a, d, a^2, d^2 or ad, contradicting the fact that $|a|,|d| > 2$ or that G is not abelian-by-finite. The only possible way for $\tilde{D}$ to contain a 2-gon is if there had been two adjacent faces in D sharing the same vertices all of which, apart from two of them, had been removed in forming $\tilde{D}$. Any attempt at labelling two such faces shows that this is impossible. A similar argument works for the case of possible 3-gons.

It can be shown, similarly to Lemma 2.5 of [6], that Lemma 2.2 forces $\tilde{D}$ to contain an interior face, P say, of positive curvature. Observe that since each $m_i \ge 3$ in (*), P must have degree m = 4 or 5. We show that in both cases a contradiction is obtained.

Lemma 2.3. *The interior vertices of degree* 3 *in* $\tilde{D}$ *are labelled by* $(a\omega)^{\pm 3}$ *or* $(\omega d)^{\pm 3}$.

Proof. For if V is an interior vertex of degree 3 and the label of V involves both a and d then the relator obtained is a subword of ad^2 or a^2d, and in each case a contradiction is obtained.

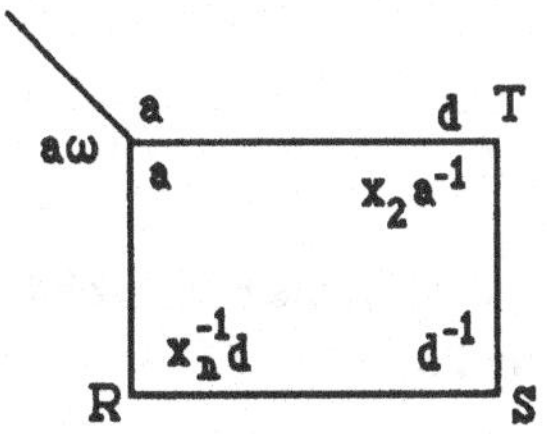

Fig. 2.4

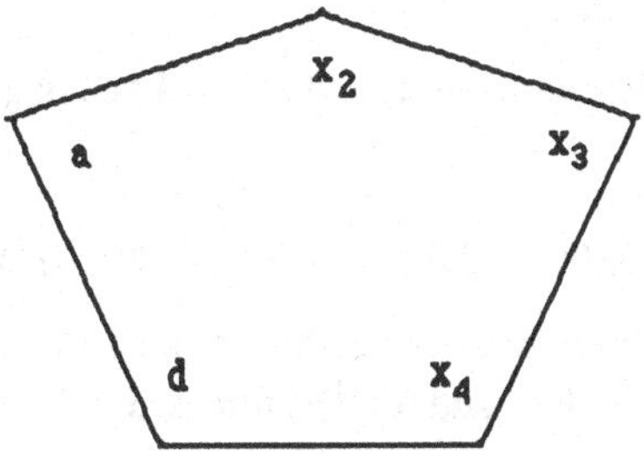

Fig. 2.5

Suppose that P is a 4-gon. Then, since $n > 3$, P has had a diagonal (that is, a lane connecting a pair of opposite corners of P) removed in forming $\tilde{D}$ from D. In fact, since we know that P must contain at least one vertex of degree 3, it follows from Lemma 2.3 that it can be assumed without any loss that P is given by Fig. 2.4, whence $a^3 = 1$ in G. We investigate the possible degrees of the other vertices R, S and T of P. Observe immediately that any occurrence of ω in Lemma 2.3 must have even length, and so deg R, deg T > 3.

If deg S = 3 then $d^3 = 1$ in G. Moreover the label of T involves both a and d, and it is easy to check that if deg T ≤ 5 then the relator obtained, together with a^3 and d^3, forces G to be residually finite. By symmetry the same holds for deg R and P cannot have positive curvature. We can assume then that deg S > 3.

If deg T = 4 then by inspection the only possible label that does not imply G is abelian-by-finite is

$$(x_2a^{-1})(\omega d)(\omega^{-1}a^{-1})(\omega d),$$

whence $(da^{-1})^2 = 1$ in G. If now deg S = 4 then S has label either $(\omega)(d^{-1})\lambda\mu$ or $(d^{-1})(d^{-1})\lambda\mu$ (where λ,μ denote basic sublabels), and in both cases G is either forced to be cyclic or we obtain the relator d^3 or d^4, each of which together with a^3 and $(da^{-1})^2$ means that G is finite.

If deg R = 4 then R has label one of

$$(x_n^{-1}d)(\omega)\lambda\mu,\ (x_n^{-1}d)(\omega d)\lambda\mu \text{ or } (x_n^{-1}d)(\omega a^{-1})\lambda\mu.$$

Clearly the relator obtained from the first of these labels will force G to be abelian-by-finite, a contradiction; from the second label we obtain the relator d^3 which, as noted above, means that deg T ≥ 6 and P cannot have positive curvature; for the third of the labels the only allowable relator is $(da^{-1})^2$, in which case μ is (ωa^{-1}). If now deg S = 4 then S must have label $(d^{-1})(d^{-1}\omega)\lambda\mu$ and, as above, we get a contradiction.

We are left with deg R = deg T = 4 and deg S = 5. But the label of S will be one of

$$\{(a\omega),(\omega)\}(d^{-1})\{(\omega),(d^{-1}\omega),(d^{-1}\omega a^{-1})\}\lambda\mu$$

and it is routine to check that for all possible choices of λ,μ the relator obtained, together with a^3 and $(da^{-1})^2$, forces G to be abelian-by-finite.

To complete this case we deal with when m = 5. The only way P can be a 5-gon is if n = 4 and P is given by Fig. 2.5. Since P has positive curvature it must have at least four vertices of degree 3, which clearly is not possible.

2.2. 2 = |a| ≤ |d|

Since G is not residually finite we can assume that |a| < |d|. The fact that D can now contain vertices of degree 2 labelled by $a^{\pm 2}$ provides the next observation.

Lemma 2.4. D *contains no vertex of degree greater than* 2 *having*

$$(axa)^{\pm 1}$$

as a sublabel, where x = 1 *in* G.

Proof. For if V were such a vetex then by means of two obvious diamond moves we can obtain a new counter-example with the same number of faces but with an extra vertex of degree 2 labelled by $a^{\pm 2}$, thus increasing N_1, a contradiction.

Remove from D all interior vertices of degree 2, other than those labelled by $a^{\pm 2}$, together with the edges incident to them. Call this new diagram D'. Remove from D' those interior vertices of degree 2 that are both labelled by $a^{\pm 2}$ and form part of some 4-gon in D'. The only way such vertices arise is given by Fig. 2.6. Observe that we have at no stage created any new vertices of degree 2 labelled by $a^{\pm 2}$. Now replace each vertex of degree 2 labelled by $a^{\pm 2}$ that remains, together with the pair of edges incident to it, by a line segment. Finally obtain $\tilde{D}$ by re-introducing line segments to take account of isolated subdiagrams consisting of more than a single vertex, as in the case of Section 2.1.

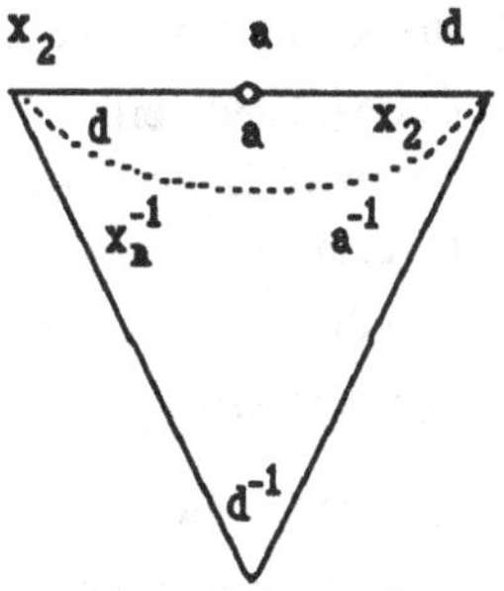

Fig. 2.6

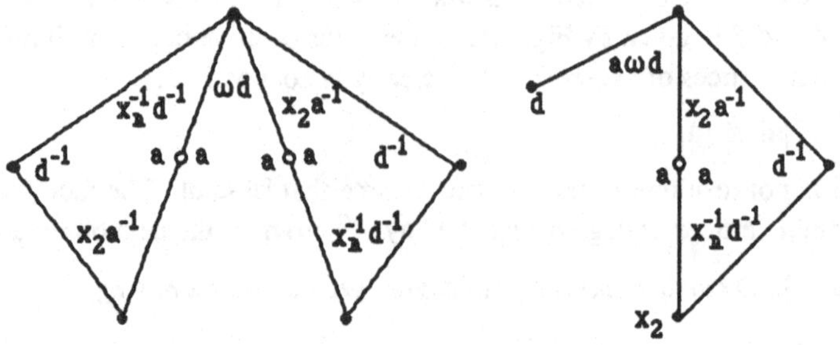

(i) (ii)

Fig. 2.7

In this case both a and d can occur twice in a sublabel of a vertex label of $\tilde{D}$ (although not the same sublabel) - see Fig. 2.7. In fact it is a straightforward check that the basic list of sublabels of $\tilde{D}$ is:

$$(a\omega dx_2a^{-1}),\ (\omega dx_2a^{-1}),\ (x_n^{-1}d\omega dx_2a^{-1}),\ (x_n^{-1}d\omega d),\ (x_n^{-1}d\omega)$$

together with their inverses and any subword of $(a\omega d)^{\pm 1}$.

Lemma 2.5. *Each interior vetex of* $\tilde{D}$ *has degree* 3 *and each face has degree at least* 4.

Proof. As in the proof of Lemma 2.2, any attempt at labelling shows that the diagram D' contains no faces of degree less than four. Furthermore, since the offending vertices of degree 2 labelled by $a^{\pm 2}$ have been removed, we do not create any 3-gon when obtaining $\tilde{D}$ from D'. This deals with the latter part of the assertion. For the first part observe that $\tilde{D}$ contains no isolated vertices. For if V were such a vertex then clearly the label of V as a vertex of D must involve the x_j $(2 \leq j \leq n)$ and d only. But the only way edges incident to vertices of D whose label involves d are removed in forming $\tilde{D}$ is essentially given by Fig. 2.6 and Fig. 2.7(i), and observe that in both cases the edge between x_n^{-1} and d^{-1} must remain. It is clear from the basic list of sublabels that if $\tilde{D}$ has an interior vertex of degree 1 then G would then be cyclic. If $\tilde{D}$ contained an interior vertex V of degree 2 then the label of V is $\lambda\mu$ where λ and μ are basic sublabels. Taking Lemma 2.4 into account it is a routine check that the relator obtained is a non-empty subword of $(ad)^2$ or $(ad^2)^2$, and in each case, since $a^2 = 1$, G is abelian-by-finite, a contradiction.

As in the previous case it follows from Lemma 2.5 that $\tilde{D}$ must contain an interior face P of positive curvature and also that P has degree 4 or 5. If P is a 4-gon then P is one of those given in Fig. 2.8.

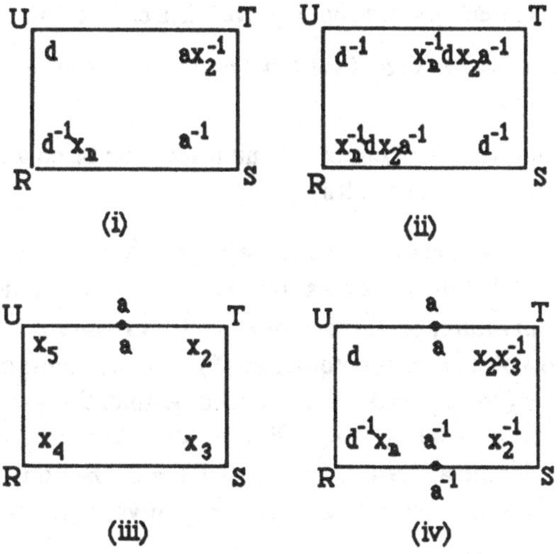

Fig. 2.8

Let V be a vertex of P and suppose that deg V = 3. Let the label of V be $\lambda\mu\nu$ where λ denotes a corner label of Fig. 2.8, and μ, ν are from the basic list of sublabels.

Lemma 2.6. *If* $|d| > 3$ *in* G *then* λ *is* $(x_n^{-1}dx_2a^{-1})$ *and* $\mu\nu$ *is either*

$$(\omega dx_2a^{-1})^2 \textit{ or } (x_n^{-1}d\omega)(a\omega dx_2a^{-1});$$

if $|d| = 3$ *then either* λ *is* $d^{\pm1}$ *and* $\mu\nu$ *is*

$$(\omega d)^{\pm2} \textit{ or } ((\omega)(x_n^{-1}d\omega d))^{\pm1},$$

or λ *is* ω $(\neq x_4^{\pm1}$ or $(x_2x_3^{-1})^{\pm1})$ *and* $\mu\nu$ *is*

$$((x_n^{-1}d\omega d)(\omega d))^{\pm1} \textit{ or } ((x_n^{-1}d\omega)(x_n^{-1}d\omega d))^{\pm1}.$$

Proof. The highest power of d that can be obtained from $\lambda\mu\nu$ is clearly d^3, whence if $|d| > 3$ then $\lambda\mu\nu$ must involve a, and so must involve both a and d at least twice. Since the relators $adad^{\ell}$ ($\ell \in Z$), $(ad^2)^2$, $(ad^2)^2ad$ and $(ad)^3d$ each together with a^2 give a contradiction to G not residually finite it is routine to check that the only allowable relator is $(ad)^3$, and that the only possible labels are the ones listed. If $|d| = 3$ and $\lambda\mu\nu$ does not involve a then the list given is easily verified. The fact that $\lambda\mu\nu$ does not involve a is a consequence of the next result.

Lemma 2.7. *If* $d^3 = 1$ *in* G *and the label of an interior vertex,* V *say, involves* a, *then* deg V > 6.

Proof. This follows from Lemma 2.4, an inspection of how the basic sublabels can be arranged, and Lemma 1.2(ii).

Let P be an interior 5-gon of $\tilde{D}$. A routine check shows that P has a corner label $d^{\pm 1}$ and a corner label ω. It follows from Lemma 2.6 that if $|d| > 3$ then P cannot have positive curvature, so assume $|d| = 3$. By Lemma 2.7 P must not have a corner label involving a, for equation (*) will be non-positive. The only possibility for P (up to inversion and cyclic permutation) is given by Fig. 2.9. Since it has a corner label x_5 (= x_n). We see (as in Lemma 2.6) that the vertex S of Fig. 2.9 has degree at least 4 whence deg T = 3. But this forces the label of vertex W to involve a, whence P cannot have positive curvature.

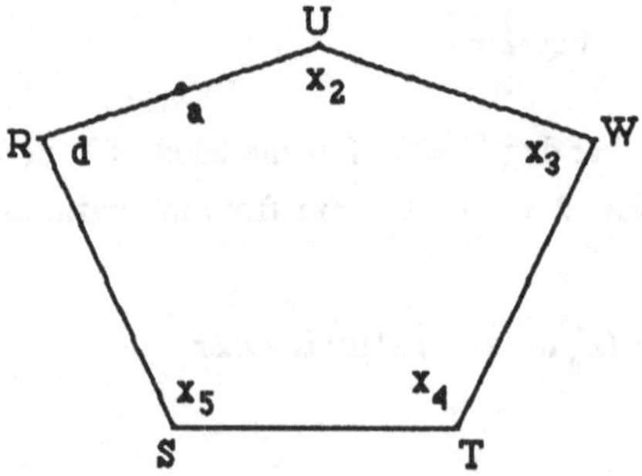

Fig. 2.9

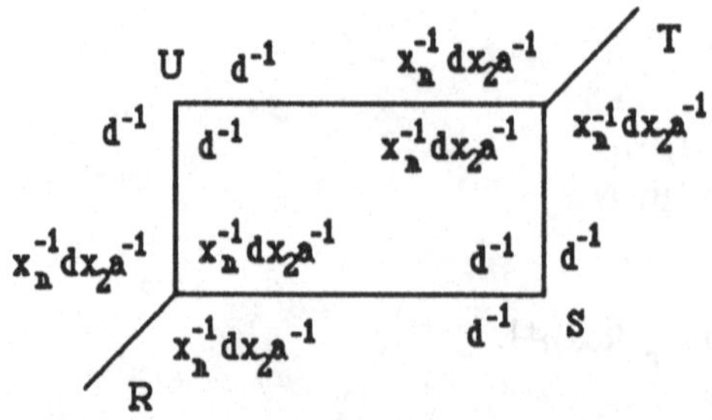

Fig. 2.10

Now let P be an interior 4-gon of $\tilde{D}$. We treat each of the four subcases of Fig. 2.8 in turn.

Subcase (i): By Lemma 2.6 the only vertex in Fig. 2.8(i) that can have degree 3 is U, whence $d^3 = 1$ in G. This also forces the label of R to involve a. Since we now have that the label of each of R, S and T involves a, it follows from Lemma 2.7 that P cannot have positive curvature.

Subcase (ii): If $d^3 = 1$ in G then, by Lemma 2.7, R and T must have degree at least 7 and P cannot have positive curvature. Thus $|d| > 3$ and we can assume that deg T = 3, so T has label $(x_n^{-1}dx_2a^{-1})^3$. If the degree of U or S is four then a routine check shows that the relator obtained either involves a once, or is one of the forms

$$adad^m, ad^2ad^q, d^q \ (|q| \leq 4),$$

and in all cases G is then abelian-by-finite. Thus deg R = 3 and P is given by Fig. 2.10. Suppose deg R = 5, that is, the label of R is either

$$(d^{-1})(d^{-1})(d^{-1}\ldots)\lambda\mu \text{ or } (ax_2^{-1}d^{-1})(d^{-1})(d^{-1}\ldots)\lambda\mu,$$

where λ and μ are basic sublabels. If a is not involved in the label then the highest power of d that can be obtained is d^6 which, together with a^2 and $(ad)^3$, forces G to be abelian-by-finite; if a is involved at least twice then the label is one of the following:

$$adad^m, ad^2ad^q \ (|q| < 6), adad^{\pm 1}ad^3, adad^{-2}ad^{\pm 3}, adad^3ad^2.$$

In all cases we can use the fact that $(ad)^3 = 1$ and Lemma 1.2 to show that G is abelian-by-finite and so deg R, and deg S by symmetry, is at least six, and P cannot have positive curvature.

Subcase (iii): Observe immediately that since at least one of U, R, S or T must have degree 3, we can conclude from Lemma 2.6 that $|d| = 3$. It can be further observed from Lemma 2.6 that deg R > 3. If deg R = 4 then a routine check shows that the label of R is either

$$(x_4)(\omega d)(\omega)(x_4^{-1}d\omega d) \text{ or } (x_4)(\omega d)^3.$$

In either case this means that the label of S involves a whence, by Lemma 2.7, deg S > 6. For P to now have positive curvature we must have deg U = deg T = 3. But, by Lemma 2.6, if deg U = 3 then the label of R must involve a, a contradiction. We can assume therefore that deg R > 4.

Assume for the moment that deg U > 3. If deg S = 3 then by Lemma 2.6 the label of T must involve a and deg T > 6, whence P cannot have positive curvature. If deg T = 3 then S must have sublabel $(d^{-1})(x_3)$, and it is routine to check that this forces deg S > 4, again a contradiction.

In conclusion we must have deg U = 3, whence deg R > 6. By the above paragraph we see that the only possibility for P to have positive curvature is if deg T = 3, deg S = 5 and deg R = 7. But since (x_4) is a sublabel of the label of R it is clear that the relator obtained will be an alternating word of length at most twelve in a and d, and Lemma 1.2(ii) gives the contradiction.

Subcase (iv): It is clear that by Lemma 2.6 deg T and deg R > 3. If deg U = 3 then |d| = 3 and the label of R involves a. So for P to have positive curvature deg S must equal 3, and the label of S is either

$$(x_2^{-1})(d^{-1}\omega)(d^{-1}\omega d^{-1}x_n) \text{ or } (x_2^{-1})(d^{-1}\omega d^{-1}x_n)(\omega d^{-1}x_n).$$

With both labels there are in fact examples of 4-gons with positive curvature. The maximum curvature that can be obtained is $+\frac{1}{6}\pi$, that is, when deg U = deg R = deg T = 4. (An example is illustrated in Fig. 2.11.)

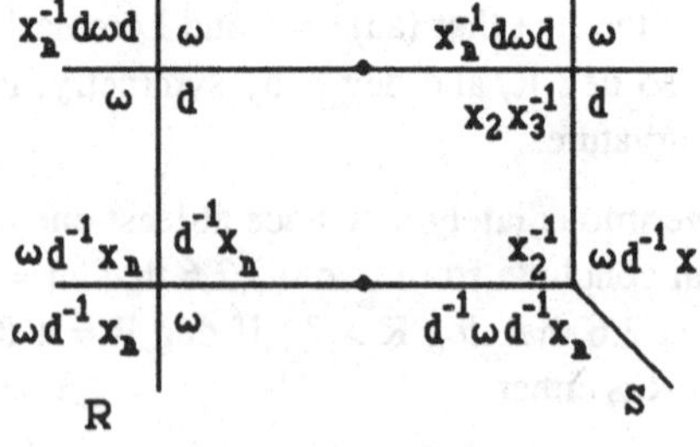

Fig. 2.11

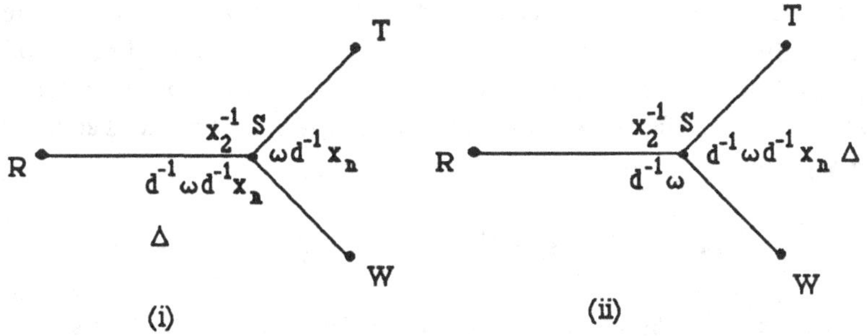

Fig. 2.12

In conclusion, each interior positively curved faced is a 4-gon and has a vertex such as the vertex S of Fig. 2.12(i) and (ii). (Thus $|d| = 3$ in G.) We shall show that these positively curved faces P are compensated for by neighbouring negatively curved faces of $\tilde{D}$.

We call P an *associated face* of the face Δ in Fig. 2.12(i) and (ii). (Thus there is a 1-1 correspondence between interior faces of $\tilde{D}$ of positive curvature and associated faces.) (For ease of presentation, associated faces shall sometimes be referred to as associates.) Since Δ has corner label $(d^{-1}\omega d^{-1}x_n)^{\pm 1}$ we see from Fig. 2.8 that $\deg \Delta > 4$. In fact, it is readily verified that Δ cannot be a 5-gon and so $\deg \Delta \geq 6$. As a further consequence of Δ having corner label $(d^{-1}\omega d^{-1}x_n)^{\pm 1}$ it can be seen from Fig. 2.7(i) that Δ also has a vertex V, say, with corner label $(ax_2^{-1}\ldots)^{\pm 1}$. If P_1 and P_2 are associated faces of Δ then the only way that two such vertices V_1 and V_2 can coincide is if Δ has corner label $(d^{-1}\omega d^{-1}x_n)^{\pm 1}$ at P_1 and $(d^{-1}\omega d^{-1}x_n)^{\mp 1}$ at P_2. But $V_1 = V_2$ must then have a occurring twice in its corner label at Δ, and this corner label as a label of D must have involved at least two $a^{\pm 2}$ segments (a lane containing exactly one intermediate vertex, and this vertex has label $a^{\pm 2}$) removed in forming $\tilde{D}$. We see from Fig. 2.7(ii) that this is impossible.

Let Δ be an interior face of $\tilde{D}$ of degree k which has $k_1 \geq 1$ associated faces. It follows from the above paragraphs that the total curvature of Δ plus its associated faces is at most

$$-(k-2)\pi + 2\pi\left(\frac{k_1}{7} + \frac{k-k_1}{3}\right) + \frac{k_1\pi}{6} \qquad (**)$$

and this value exceeds zero precisely when $2 > \frac{1}{3}k + \frac{3}{14}k_1$. Since $k \geq 6$ and $k_1 \geq 1$, this can never happen.

Now let Δ be a non-interior face of $\tilde{D}$ which has $k_1 \geq 1$ associates. The greatest positive difference that can be made to (**) is to replace one of the vertices of degree ≥ 7 with the distinguished vertex V_0. This way the total curvature of V_0 plus its associated faces (which are interior, so remain unaffected) is at most (with $\deg V_0 = \lambda$)

$$-(k-2)\pi + 2\pi\left(\frac{k_1-1}{7} + \frac{1}{\lambda} + \frac{k-k_1}{3}\right) + \frac{k_1\pi}{6} = \pi\left(\frac{12}{7} + \frac{2}{\lambda} - \frac{k}{3} - \frac{3k_1}{14}\right).$$

The greatest value this can achieve is with $k = 6$ and $k_1 = 1$ which gives

$$\pi(2/\lambda - 1/2).$$

So for a positive total curvature we require either $\deg V_0 = 3$ to obtain $\frac{1}{6}\pi$ or $\deg V_0 = 2$ to obtain $\frac{1}{2}\pi$.

Since we have shown that if Δ is an interior face of $\tilde{D}$ then the total curvature of Δ plus its associated faces is non-positive, we deduce that the total curvature C, say, of the non-interior faces of $\tilde{D}$ together with all of their associated faces is at least 4π. We finish the proof of the theorem by showing that this is impossible.

Now for $\deg V_0 = \lambda$ the greatest possible curvature a non-interior face can have is $2\pi/\lambda$ (obtained from a 4-gon all of whose interior vertices have degree 3) and we see that if $\lambda = 3$ then this value, $\frac{2}{3}\pi$, exceeds $\frac{1}{6}\pi$, and if $\lambda = 2$ the value, π, exceeds $\frac{1}{2}\pi$. This means that the total curvature C is at most $\lambda(2\pi/\lambda) = 2\pi < 4\pi$, our final contradiction.

References

1. D J Collins & J Huebschmann, Spherical diagrams and identities among relations, *Math. Ann.* **261** (1982), 155-183.
2. M Conder, Three-relator quotients of the modular group, *Quart. J. Math. Oxford* (2) **38** (1987), 427-447.
3. M Edjvet, Equations over groups and a theorem of Higman, Neumann and Neumann, *Proc. London Math. Soc.*, to appear.
4. M Edjvet & J Howie, The solution of length four equations over groups, *Trans. Amer. Math. Soc.*, to appear.
5. S M Gersten, Products of conjugacy classes in a free group: a counter-example, *Math. Z.* **192** (1986), 167-181.
6. J Howie, The solution of length three equations over groups, *Proc. Edinburgh Math. Soc.* **26** (1983), 89-96.

7. F Levin, Solutions of equations over groups, *Bull. Amer. Math. Soc.* **68** (1962), 603-604.

8. O S Rothaus, On the nontriviality of some group extensions given by generators and relations, *Ann. of Math.* **106** (1977), 559-612.

GENERALIZING ALGEBRAIC PROPERTIES OF FUCHSIAN GROUPS

BENJAMIN FINE

Fairfield University, Fairfield, Connecticut 06430, USA

GERHARD ROSENBERGER

Universität Dortmund, 4600 Dortmund 50, Federal Republic of Germany

Dedicated to the memory of Horace Mochizuki

1. Introduction

What we wish to describe is a program for generalizing Fuchsian Groups. Recall that a *Fuchsian group* is a discrete subgroup of $PSL_2(\mathbf{R})$ or a conjugate of such a group in $PSL_2(\mathbf{C})$. If F is finitely generated then F has a standard Poincaré presentation of the following form:

$$F = < e_1,...,e_p,h_1,...,h_t,a_1,b_1,...,a_g,b_g \,;\, e_i^{m_i} = 1,\ i = 1,...,p,\ R = 1 >$$

$$\text{where } R = UV \text{ with } U = e_1 \ldots e_p h_1 \ldots h_t,\ V = [a_1,b_1] \ldots [a_g,b_g], \qquad (1)$$

$$p \geq 0,\ t \geq 0,\ g \geq 0,\ p+t+g > 0, \text{ and } m_i \geq 2 \text{ for } i = 1,...,p.$$

{U or V could possibly be trivial}.

A group G with a presentation of form (1) actually represents a Fuchsian group if

$$\mu(G) = 2g - 2 + t + \Sigma(1 - 1/m_i) > 0.$$

$\mu(G)$ represents the hyperbolic area of a fundamental polygon for G, [5]. We call groups with presentation (1) F-*groups*. If $t \neq 0$ then F is a free product of cyclics so we include all finitely generated free products of cyclics among the F-groups. Further if $p = t = 0$ then F is a surface group of genus g so all surface groups are included among the F-groups. A group-theoretical discussion of F-groups is given in Lyndon and Schupp's book [14].

To generalize F-groups we place them in the wider context of one-relator products of cyclics. If $\{G_\alpha\}$, $\alpha \in Å$, is a class of groups then a one-relator product is a group of the form $G = (*G_\alpha)/N(R)$ where R is a non-trivial element in the free product $*G_\alpha$. The groups G_α are called the *factors* and R is the *relator*. In this context a one-relator group is a one-relator product of free groups and an F-group is a one-relator product of cyclic groups. In the latter case $G_\alpha = \langle g_i \rangle$ where g_i is a generator and R = UV. In a series of papers [8,9,10,11,12], Fine, Rosenberger, Howie and Levin developed a comprehensive theory of one-relator products of cyclics where the relator is a proper power - that is groups of the form

$$G = \langle x_1,\ldots,x_n \, ; \, x_i^{m_i} = 1, i = 1,\ldots,n, S^m = 1 \rangle \qquad (2)$$

where $m_i = 0$ or $m_i \geq 2$ for $i = 1,\ldots,n$, S is a cyclically reduced word in the free product on $\{x_1,\ldots,x_n\}$ and $m \geq 2$. (Further more recent work on this has been done by Tang [28].)

This theory showed that there is a Freiheitssatz for such groups and further these groups share many properties with linear groups. In particular in most cases they satisfy the Tits Alternative - that is either they contain free subgroups of rank 2 or are virtually solvable. Further many are virtually torsion-free and decompose as non-trivial amalgams. A description of all these results can be found in [9].

In the present program we try to generalize F-groups by considering the case of one-relator products of cyclics where the relator is not a proper power. Essentially we consider the following general question:

Question: Given an algebraic property of an F-group, how does it extend (if at all) to general one-relator products of cyclics -

$$\langle x_1,\ldots,x_n \, ; \, x_i^{m_i} = 1, i = 1,\ldots,n, R_1 = 1 \rangle?$$

In particular what conditions on the relator R_1 allow a particular result on F-groups to go through?

Our work to date indicates that if $n \geq 3$ and $R_1 = U_1V_1$ with

$$U_1 = U_1(x_1,\ldots,x_p),\ V_1 = V_1(x_{p+1},\ldots,x_n),\ p \geq 1$$

and both U_1 and V_1 not proper powers and of infinite order in the free products on the respective generators they involve then many results on F-groups carry through.

2. Notation and representation results

We define *a group of F-type* to be a group which has a presentation of the following form

$$G = < a_1,...,a_n \,;\, a_1^{e_1} = ... = a_n^{e_n} = 1,\ U(a_1,...,a_p)V(a_{p+1},...,a_n) = 1 > \qquad (3)$$

where $n \geq 2$, $e_i = 0$ or $e_i \geq 2$, $1 \leq p \leq n-1$, $U(a_1,...,a_p)$ is a cyclically reduced word in the free product on $a_1,...,a_p$ which is of infinite order and $V(a_{p+1},...,a_n)$ is a cyclically reduced word in the free product on $a_{p+1},...,a_n$ which is of infinite order. With p understood we write U for $U(a_1,...,a_p)$ and V for $V(a_{p+1},...,a_n)$.

Now if $U = a_1^{\pm 1}$ then e_1 must equal zero since we assume that U has infinite order. Thus in this case G reduces to

$$G = < a_2,...,a_n \,;\, a_2^{e_2} = ... = a_n^{e_n} = 1 >$$

which is a free product of cyclics and thus an F-group. Therefore if $p = 1$ (or $p = n-1$) we restrict groups of F-type to those where $U = a_1^m$ (or $V = a_n^m$) with $|m| \geq 2$. It follows then that in all cases the group G decomposes as a non-trivial free product with amalgamation:

$$G = G_1 \underset{A}{*} G_2 \qquad (4)$$

where the factors are free products of cyclics

$$G_1 = < a_1,...,a_p \,;\, a_1^{e_1} = ... = a_p^{e_p} = 1 >$$

$$G_2 = < a_{p+1},...,a_n \,;\, a_{p+1}^{e_{p+1}} = ... = a_n^{e_n} = 1 >$$

and

$$A = <U^{-1}> = <V>.$$

Recall that an F-group which cannot be generated by two or less elements always has a presentation as a group of F-type - with U and V possibly different from those in (1).

Our basic program is then to extend results on F-groups to groups of F-type. The extensions of such results depend on two facts; first the free product with amalgamation decomposition given above and second the fact that groups of F-type can be represented (sometimes faithfully) in $PSL_2(C)$.

In [8], Fine, Howie and Rosenberger using a refinement of a technique in Baumslag, Morgan and Shalen [4] showed that if G is a group of form (2) - that is a one-relator product of cyclics where the relator S^m is a proper power such that S involves all the generators then there exists a representation

$$\rho: G \to PSL_2(C)$$

such that $\rho(a_i)$ has order e_i if $e_i \geq 2$ and infinite order if $e_i = 0$ and $\rho(S)$ has order m. Such a representation is called an ***essential representation***. The existence of these essential representations was used to prove a Freiheitssatz for groups of form (2). That is if S involves all the generators the subgroup generated by any proper subset of the generators is the obvious free product of cyclics. In the language of Tang [28] if $m \geq 2$ then any relator S^m is a *Magnus relator*. (See [28] for terminology.) The above representation result was extended to the following case:

Suppose A and B are groups which admit faithful representations in $PSL_2(C)$ and R is a cyclically reduced element of the free product A*B with syllable length at least two and suppose that $m \geq 2$. Then the group

$$G = (A*B)/N(R^m)$$

admits a representation in $PSL_2(C)$ such that $\rho|_A$ is faithful, $\rho|_B$ is faithful and $\rho(R)$ has order m. In particular both A and B inject into G. (These injections constitute the Freiheitssatz for such one-relator products.)

In general if m = 1 there do not necessarily exist essential representations in $PSL_2(C)$ [8]. However if R = UV as in the groups of F-type there are essential representations and further if neither U nor V is a proper power there exist faithful representations.

Theorem 1. *Let* G *be a group of* F*-type. Then* G *admits a representation* $\rho: G \to PSL_2(C)$ *such that* $\rho|_{G_1}$ and $\rho|_{G_2}$ *are faithful* (G_1, G_2 *the factors of* G *from* (4)). *Further if neither* U *nor* V *is a proper power then* G *has a faithful representation in* $PSL_2(C)$.

Proof. First suppose that UV omits some generator. For instance suppose that UV does not involve a_1. Then G is a free product $H_1 * H_2$ where

$$H_1 = < a_1 \,;\, a_1^{e_1} = 1 >$$

$$H_2 = < a_2, \ldots, a_n \,;\, a_2^{e_2} = \ldots = a_n^{e_n} = UV = 1 >.$$

If H_2 admits a representation $\phi : H_2 \to PSL_2(C)$ such that

$$< a_2, \ldots, a_p \,;\, a_2^{e_2} = \ldots = a_p^{e_p} = 1 > \to H_2 \xrightarrow{\phi} PSL_2(C)$$

and

$$G_2 \to H_2 \xrightarrow{\phi} PSL_2(C)$$

are faithful then G has a representation $\rho: G \to PSL_2(C)$ such that $\rho|_{G_1}$ and $\rho|_{G_2}$ are faithful (see [8]). Therefore we can assume that UV involves all the generators.

Choose faithful representations

$$\sigma_1 : G_1 \to PSL_2(C)$$

$$\sigma_2 : G_2 \to PSL_2(C)$$

such that

$$\sigma_1(U^{-1}) = \begin{bmatrix} t_1 & 0 \\ 0 & t_1^{-1} \end{bmatrix} \text{ and } \sigma_2(V) = \begin{bmatrix} t_2 & 0 \\ 0 & t_2^{-1} \end{bmatrix}$$

where t_1 and t_2 are transcendental over **Q**. This may be done since U and V both have infinite order and if G_i, $i = 1$ or 2, is non-cyclic then the dimension of the character space as an affine variety is positive [11].

From work of Shalen (Lemma 3.2 of [26]) there exists an automorphism α of C such that $\alpha(t_1) = t_2$. We define a faithful representation ρ_1 of G_1 by

$$\rho_{1ij}(g) = \alpha(\sigma_{1ij}(g)),\ i = 1,2 \text{ and } j = 1,2,$$

where $\rho_{1ij}(g)$ (respectively $\sigma_{1ij}(g)$) is the *ijth* entry of $\rho_1(g)$ (respectively of $\sigma_1(g)$). Further let $\rho_2 = \sigma_2$. Then ρ_2 is faithful on G_2. G_1 and G_2 generate G and let ρ be the representation of G induced by ρ_1 and ρ_2. This gives the desired representation of the theorem. Further if neither U nor V is a proper power the above construction leads to the existence of a faithful representation of G (see [20]).

We note that if both U and V are proper powers then there is no faithful representation in $PSL_2(C)$. If $U = U_1^\alpha$, $\alpha \geq 2$ and $V = V_1^\beta$, $\beta \geq 2$ and $\rho: G \to PSL_2(C)$ is a representation and $\rho(U)$ and $\rho(V)$ have infinite order then $\rho(U_1)$ and $\rho(V_1)$ must commute. However U_1 and V_1 do not commute in G. (Non-elliptic elements of $PSL_2(C)$ commute if and only if they have the same fixed points [15]. Therefore $\rho(U_1^\alpha)$ commutes with $\rho(V_1^\beta)$ implies that $\rho(U_1)$ commutes with $\rho(V_1)$.)

There are several immediate consequences of Theorem 1 which we describe in the next section. We note that using exactly the same argument the following generalization of Theorem 1 can be obtained.

Theorem 2. (1) *Let* H_1 *and* H_2 *be groups and* U_1, U_2 *elements of infinite order in* H_1, H_2, *respectively. Suppose that each* H_i, $i = 1,2$, *admits a faithful representation* ρ_i *in* $PSL_2(C)$ *such that* $tr(\rho_i(U_i))$ *is transcendental. Let*

$$H = (H_1 * H_2)/N(U_1U_2)$$

be the one-relator product of H_1, H_2 *with relator* U_1U_2. *Then* H *admits a representation* $\rho: H \to PSL_2(C)$ *such that* $\rho|H_1$ *and* $\rho|H_2$ *are faithful.*

(2) *Suppose* H *is as in* (1) *but assume further that each solvable subgroup of* H_i, $i = 1,2$ *is either cyclic or infinite dihedral and* U_i *is not a proper power in* H_i. *Then* H *admits a faithful representation in* $PSL_2(C)$.

Many of the consequences which follow for groups of F-type from Theorem 1 will also hold for groups H of the form of Theorem 2. We will point these out also in the next section.

3. Algebraic consequences

The well-known Fenchel-Fox Theorem [14] aserts that a finitely generated Fuchsian group is virtually torsion-free - that is contains a torsion-free subgroup of finite index. This has been extended by Selberg [25] to general finitely generated linear groups. In [11], Fine, Levin and Rosenberger showed that many one-relator products of cyclics with proper power relators are virtually torsion-free. Further since Fuchsian groups can be represented as groups of projective matrices it follows that the finitely generated Fuchsian groups are residually finite and thus Hopfian [14]. Using Theorem 1 these two results, virtual torsion-free property and residual finiteness, carry over to many groups of F-type.

Theorem 3. *Let* G *be a group of* F*-type. Then:*

(1) G *is virtually torsion-free.*

(2) *If neither* U *nor* V *is a proper power then* G *is residually finite and thus Hopfian.*

Proof. Let ρ:G $\to PSL_2(C)$ be a representation of G in $PSL_2(C)$ constructed as in Theorem 1. Then ρ(G) is a finitely generated subgroup of $PSL_2(C)$. From the results of Selberg ρ(G) contains a normal torsion-free subgroup H* of finite index. This pulls back to a normal subgroup H of G of finite index. We claim H is torsion-free. Let g $\in$ G be a non-trivial element of finite order. Since G is a free product with amalgamation g is conjugate to an element of finite order in one of the factors. Since each factor is a free product of cyclics g must be conjugate to a power of one of the generators a_i. The representation ρ is faithful on each factor so it follows that $\rho(a_i)$ has exactly the same order in ρ(G) as a_i has in G. Therefore ρ(g) must have finite order ≥ 2 in ρ(G) and thus ρ(G) is not in H*. It follows that g is not in H and thus H is torsion-free.

If neither U nor V is a proper power then G has a faithful representation ρ in $PSL_2(C)$. Therefore ρ(G) is a finitely generated subgroup of $PSL_2(C)$ and is thus residually finite [14]. It follows that G is residually finite. From a theorem of Malcev [17] this implies that G is Hopfian.

Through the representation in $PSL_2(C)$ these results can be extended to the more general one-relator products of Theorem 2.

Theorem 4. *Let* H *be a one-relator product of the form of Theorem 2. Then* H *is virtually torsion-free. If in addition* H *satisfies the conditions of (ii) of Theorem 2 then* H *is residually finite and Hopfian.*

Since a group of F-type decomposes as a free product with amalgamation this decomposition can be used to give a great deal of information about the finite subgroups. In particular each finite subgroup is cyclic corresponding to the same statement for Fuchsian groups.

Theorem 5. *Let* G *be a group of* F-*type with presentation* (3). *Then:*

(1) *If* $e_i \geq 2$ *then* a_i *has order exactly* e_i.

(2) *Any element of finite order in* G *is conjugate to a power of some* a_i.

(3) *Any finite subgroup is cyclic and conjugate to a subgroup of some* $<a_i>$.

Proof. The proofs follow directly from the decomposition of G as a free product with amalgamation and the corresponding properties of generalized free products. Any element of finite order and any finite subgroup in G must be conjugate to an element of finite order or a finite subgroup contained in one of the factors. Since each factor in a group of F-type is a free product of cyclics it follows that any finite subgroup is cyclic. Further since the finite subgroups in a free product of cyclics are precisely the conjugates of the cyclic subgroups generated by the generators the other statements follow.

Theorem 1 says that a group of F-type is "almost" a finitely generated matrix group. Finitely generated matrix groups satisfy the Tits alternative - that is either they contain a free subgroup of rank 2 or they are virtually solvable. In [10], Fine, Levin and Rosenberger showed that almost all one-relator products of cyclics with proper power relators (form (2)) satisfy the Tits alternative. From the free product with amalgamation decomposition the Tits alternative can be extended to groups of F-type. We have that either G has a free subgroup of rank 2 or G is solvable and restricted to one of several special cases.

Theorem 6. *Let* G *be a group of* F-*type. Then either* G *has a free subgroup of rank* 2 *or* G *is solvable and isomorphic to groups with one of the following presentations*:

(i) $H_1 = < a,b \, ; \, a^2b^2 = 1 >$

(ii) $H_2 = < a,b,c \, ; \, a^2 = b^2 = abc^2 = 1 >$

(iii) $H_3 = < a,b,c,d \, ; \, a^2 = b^2 = c^2 = d^2 = abcd = 1 >$.

Proof. Suppose $n > 4$. Then a free product of cyclics of rank ≥ 3 injects into G and therefore G has a free subgroup of rank 2. Next suppose $n = 4$ and not all $e_i = 2$. Then a free product of two cyclic groups not $\mathbf{Z}_2 * \mathbf{Z}_2$ injects into G and therefore G has a free subgroup of rank 2.

If $n = 4$ and all $e_i = 2$ then necessarily since U and V have infinite order we must have $p = 2$.

Then G has a presentation

$$< a,b,c,d \; ; \; a^2 = b^2 = c^2 = d^2 = (ab)^s(cd)^t = 1 >$$

with $s \geq 1, t \geq 1$. G then has as a factor group the free product

$$G^* = < a,b,c,d \; ; \; a^2 = b^2 = c^2 = d^2 = (ab)^s = (cd)^t = 1 >$$

$$= < a,b \; ; \; a^2 = b^2 = (ab)^s = 1 > * < c,d \; ; \; c^2 = d^2 = (cd)^t = 1 >.$$

If $(s,t) \neq (1,1)$ then G^* has a free subgroup of rank 2 and therefore G also has a free subgroup of rank 2.

If $(s,t) = (1,1)$ then G has the presentation

$< a,b,c,d \; ; \; a^2 = b^2 = c^2 = d^2 = abcd = 1 >$ which is solvable.

If $n = 3$ at least one $e_i = 0$ since U and V are assumed to have infinite order. Suppose without loss of generality that $e_1 = 0$ so that a_1 has infinite order. If $p = 2$ then a free product of two cyclic groups not $\mathbf{Z}_2 * \mathbf{Z}_2$ injects into G and therefore G has a free subgroup of rank 2. Now let $p = 1$. Then $U = a_1^{\pm s}$ with $s \geq 2$. Suppose first that $(e_2,e_3) \neq (2,2)$. Then a free product of two cyclic groups not $\mathbf{Z}_2 * \mathbf{Z}_2$ injects into G and therefore G has a free subgroup of rank 2. Next suppose that $(e_2,e_3) = (2,2)$. Then G has a presentation

$$< a,b,c \; ; \; b^2 = c^2 = a^s(bc)^t = 1 >$$

where $s \geq 2$ and $t \geq 1$. G then has as a factor group the free product

$$G^* = < a,b,c \; ; \; a^s = b^2 = c^2 = (bc)^t = 1 > =$$

$$< a \; ; \; a^s = 1 > * < b,c \; ; \; b^2 = c^2 = (bc)^t = 1 >.$$

If $s > 2$ or $t \geq 2$ then a free product of cyclics of rank ≥ 2 and not $\mathbf{Z}_2 * \mathbf{Z}_2$ injects into G^* and thus G^* has a free subgroup of rank 2. It follows that G does also. This leaves the case where $s = 2$ and $t = 1$. G then has the presentation

$$< a,b,c \; ; \; b^2 = c^2 = a^2bc = 1 >$$

which is solvable.

Finally suppose $n = 2$. Then since U and V have infinite order both generators must have infinite order and G must have a presentation

$$< a,b \; ; \; a^s b^t = 1 >$$

with $s \geq 2, t \geq 2$. G then has as a factor group the free product of cyclics

$$G^* = < a,b \; ; \; a^s = b^t = 1 >.$$

If $(s,t) \neq (2,2)$ then G^* has a free subgroup of rank 2 and therefore G does also. This leaves the situation where $(s,t) = (2,2)$. G then has the presentation

$$< a,b \; ; \; a^2 b^2 = 1 >$$

which is solvable.

If G is not solvable and therefore not isomorphic to H_1, H_2 or H_3 we can prove a stronger result.

Theorem 7. *Let* G *be a group of* F-*type. If* G *is not solvable then* G *has a subgroup of finite index which maps onto a free subgroup of rank* 2.

Proof. First assume that $n \geq 5$ or $n = 4$ and at least one $e_i \neq 2$. We may assume without loss of generality that each $e_i \geq 2$ passing to an epimorphic image if necessary.

Let $\rho : G \to PSL_2(C)$ be an essential representation constructed as in Theorem 1. It follows from the work of Selberg that there exists a normal torsion-free subgroup N of finite index in $\rho(G)$. Let π be the canonical epimorphism from $\rho(G)$ onto $\rho(G)/N$. The composition ϕ of the maps in the sequence

$$G \xrightarrow{\rho} \rho(G) \xrightarrow{\pi} \rho(G)/N$$

gives a representation of G onto a finite group. Further $\phi(a_i)$ has order e_i in $\rho(G)/N$ since ρ is essential and N is torsion-free. Now consider the free product of finite cyclic groups

$$X = < a_1,\ldots,a_n \; ; \; a_1^{e_1} = \ldots = a_n^{e_n} = 1 >.$$

There is a canonical epimorphism $\varepsilon : X \to G$. Consider the sequence

$$X \overset{\varepsilon}{\to} G \overset{\phi}{\to} \rho(G)/N.$$

Let $Y = \ker(\phi \circ \varepsilon)$. Then Y is a normal subgroup of finite index say j in X and

$$Y \cap \langle a_i \rangle = \{1\}$$

for $i = 1,\ldots,n$. Then by the Kurosh theorem Y is a free group of finite rank r. The finitely generated free product of cyclic groups X may be considered a Fuchsian group with finite hyperbolic area [5]. Therefore from the Riemann-Hurwitz formula we have

$$j\mu(X) = \mu(Y) \text{ where } \mu(X) = n - 1 - ((1/e_1) + \ldots + (1/e_n)),\ \mu(Y) = r - 1.$$

Therefore

$$r = 1 - j((1/e_1) + \ldots + (1/e_n) - n + 1).$$

The group G is obtained from X by adjoining the additional relation $UV = 1$ and thus $G = X/K$ where K is the normal closure of UV in X. Since $K \subset Y$ the factor group Y/K may be regarded as a subgroup of finite index in G. Using work of Baumslag, Morgan and Shalen (Corollary 3 of [4]) Y/K can be defined on r generators with j relations. Thus the deficiency of this presentation is given by

$$\begin{aligned} d = r - j &= 1 - j((1/e_1) + \ldots + (1/e_n) - n + 2) \\ &= 1 + j(n - 2 - (1/e_1) - \ldots - (1/e_n)). \end{aligned}$$

If $n \geq 5$ or $n = 4$ and at least one $e_i \neq 2$ then $d \geq 2$. It follows then from a theorem of B Baumslag and Pride [2] that G contains a subgroup of finite index which maps onto a free group of rank 2.

Next suppose $n = 4$ and all $e_i = 2$. Then necessarily $p = 2$ and

$$U = (a_1a_2)^s,\ V = (a_3a_4)^t \text{ with } |s| \geq 1,\ |t| \geq 1.$$

Further since G is non-solvable we must have $|s| \geq 2$ or $|t| \geq 2$. Without loss of generality we assume that $s \geq 2$ and $t \geq 1$. Then G has as a factor group the free product

$$G^* = \langle a_1, a_2, a_3 ;\ a_1^2 = a_2^2 = (a_1a_2)^s = 1,\ a_3^2 = 1 \rangle.$$

G^* has as a normal subgroup of index 2 a group isomorphic to the free product of three cyclic groups

$$\langle x, y, z ;\ x^s = y^2 = z^2 = 1 \rangle.$$

Therefore G^* and hence G also has a subgroup of finite index mapping onto a free group of rank 2.

Now suppose $n = 3$ and at least two $e_i \neq 2$. Then without loss of generality we may assume that $p = 1$, a_1 has infinite order and $U = a_1^s$ with $s \geq 2$. Suppose first that a_2 also has infinite order and $V = a_2^{\pm t}$ with $t \geq 2$. Then G has as a factor group the free product of cyclics

$$G^* = \langle a_1, a_2, a_3 ;\ a_1^s = a_2^t = a_3^{e_3} = 1 \rangle.$$

G^* and therefore G also has a subgroup of finite index mapping onto a free group of rank 2.

Now suppose that $V = a_2^{\varepsilon}$, $\varepsilon = \pm 1$. Then G is isomorphic to the free product

$$\mathbf{Z} * \mathbf{Z}_{e_3}$$

which has a subgroup of finite index mapping onto a free group of rank 2. Therefore we may now suppose that V involves both a_2 and a_3. Choose a prime number $p \geq 7$ such that p is relatively prime to s. Then G has as a factor group

$$G^* = \langle a, a_2, a_3 ;\ a^p = a_2^{e_2} = a_3^{e_3} = aV = 1 \rangle$$

$$= \langle a_2, a_3 ;\ a_2^{e_2} = a_3^{e_3} = V^p = 1 \rangle$$

with $p \geq 7$ and $(e_2, e_3) \neq (2,2)$.

From work of Rosenberger [19] (see also [4]) G^* and thus G has a subgroup of finite index mapping onto a free group of rank 2.

Next suppose $n = 3$ and that two $e_i = 2$. Then we may assume without loss of generality that a_1 has infinite order, $U = a_1^s$ with $s \geq 2$, $(e_2,e_3) = (2,2)$ and $V = (a_2a_3)^t$ with $t \geq 1$. If $t \geq 2$ then G has as a factor group the free product

$$G^* = < a_1,a_2,a_3 ;\ a_1^s = 1, a_2^2 = a_3^2 = (a_2a_3)^t = 1 >$$

which has a normal subgroup of index two isomorphic to the free product of cyclics

$$< x,y,z ;\ x^t = y^s = z^s = 1 >.$$

Hence G* and therefore also G has a subgroup of finite index mapping onto a free group of rank 2 if $t \geq 2$. If $t = 1$ then G has the presentation

$$G = < a_1,a_2,a_3 ;\ a_2^2 = a_3^2 = a_1^s a_2 a_3 = 1 >$$

$$= < a_1,a_2 ;\ a_2^2 = (a_1^s a_2)^2 = 1 >.$$

Since G is not solvable we have $s \geq 3$. Then G has as a factor group the free product of cyclics

$$< a_1,a_2 ;\ a_1^s = a_2^2 = 1 >.$$

Since $s \geq 3$ this has a subgroup of finite index mapping onto a free group of rank 2 and therefore G also has a subgroup of finite index mapping onto a free subgroup of rank 2.

Finally suppose $n = 2$. Since U and V are assumed to have infinite order and G is not solvable it follows that G must have a presentation

$$< x,y ;\ x^s y^t = 1 >$$

with $s \geq 2$ and $t \geq 3$. Then G has as a factor group the free product of cyclics

$$< x,y ;\ x^s = y^t = 1 >.$$

Since $(s,t) \neq (2,2)$ G* and hence G also has a subgroup of finite index mapping onto a free group of rank 2. This completes Theorem 5.

Since groups which map onto non-abelian free groups are SQ-universal and since a group is SQ-universal if a subgroup of finite index is SQ-universal we get the following corollary.

Corollary 1. *Let* G *be a group of* F-*type. If* G *is not solvable then* G *is* SQ-*universal.*

We note that P M Neumann originally proved that Fuchsian groups are SQ-universal [18] while J Ratcliffe [19] using arguments involving the Euler characteristic showed that every non-elementary Kleinian group is SQ-universal. G Rosenberger [21] showed that many one-relator products of cyclics are SQ-universal.

4. Subgroup theorems

One of the cornerstones of combinatorial group theory is the Freiheitssatz originally proved by Magnus which asserts that in a one-relator group in which each generator is involved in the relator any proper subset of the generators generates a free group. In general a Freiheitssatz is said to hold for a one-relator product

$$H = (A*B)/N(R)$$

if both factors A and B inject into H. Using representation results like those in Section 2, Fine, Howie and Rosenberger [8] showed that if both A and B admit faithful representations in $PSL_2(C)$, R is a cyclically reduced element of A*B of syllable length at least two and $m \geq 2$ then

$$H = (A*B)/N(R^m)$$

satisfies a Freiheitssatz. Further if an F-group is given by a standard presentation (1) it was shown in [11] that any (n-2) of the given generators generate a free product of cyclics. We next show that this is true for groups of F-type if the relator UV involves all the generators. First we need the following technical lemma which comes from work on extended Nielsen reduction in free products with amalgamation done by Zieschang [30] and Rosenberger [22].

Lemma A ([30], [22]). *Let* $G = G_1 \underset{A}{*} G_2$ *and assume that a length* L *and an order are introduced in* G *as in* [30] *and* [22]. *If* $\{x_1,\ldots,x_m\}$ *is a finite system of elements in* G *then there is a Nielsen transformation from* $\{x_1,\ldots,x_m\}$ *to a system* $\{y_1,\ldots,y_m\}$ *for which one of the following cases holds:*

(i) *Each* $w \in \langle y_1,\dots,y_m \rangle$ *can be written as*

$$\prod_{i=1}^{q} y_{v_i}^{\varepsilon_i},\ \varepsilon_i = \pm 1,\ \varepsilon_i = \varepsilon_{i+1} \text{ if } v_i = v_{i+1} \text{ with } L(y_{v_i}) \le L(w) \text{ for } i = 1,\dots,q.$$

(ii) *There is a product*

$$a = \prod_{i=1}^{q} y_{v_i}^{\varepsilon_i},\ a \neq 1, \text{ with } y_{v_i} \in A\ (i = 1,\dots,q)$$

and in one of the factors G_j *there is an element* $x \in A$ *with* $x^{-1}ax \in A$.

(iii) *Of the* y_i *there are* p, $p \ge 1$, *contained in a subgroup of* G *conjugate to* G_1 *or* G_2 *and a certain product of them is conjugate to a non-trivial element of* A.

(iv) *There is a* $g \in G$ *such that for some* $i \in \{1,\dots,m\}$ *we have* $y_i \in gAg^{-1}$, *but for a suitable natural number* k *we have* $y_i^k \in gAg^{-1}$.

The Nielsen transformation can be chosen so that $\{y_1,\dots,y_m\}$ is smaller than $\{x_1,\dots,x_m\}$ or the lengths of the elements of $\{x_1,\dots,x_m\}$ are preserved. Further if $\{x_1,\dots,x_m\}$ is a generating system of G then in the case (iii) we find that $p \ge 2$ because then conjugations determine a Nielsen transformation. If we are interested in the combinatorial description of $\langle x_1,\dots,x_n \rangle$ in terms of generators and relations we find again that $p \ge 2$ in case (iii) possibly after suitable conjugations.

Theorem 8 (Freiheitssatz for groups of F-type). *Let* G *be a group of* F-*type. If the relator* UV *involves all the generators then any subset of* (n-2) *of the given generators generates a free product of cyclics of the obvious orders.*

Proof. Let G be a group of F-type with decomposition (4) and let G_1, G_2 and p be as in the decomposition. Suppose $\{x_1,\dots,x_{n-2}\}$ is a subset of $\{a_1,\dots,a_n\}$.

Suppose first that p-1 of the x_i are elements of G_1 and n-p-1 of the x_i are elements of G_2. Assume without loss of generality that $x_i = a_i$, $i = 2,\dots,n-1$. Since UV involves all the generators it follows that $\langle a_2,\dots,a_p \rangle$ contains no element which is conjugate to a non-trivial power of U. Similarly $\langle a_{p+1},\dots,a_{n-1} \rangle$ contains no element which is conjugate to a non-trivial power of V. Therefore by Lemma A, $\langle a_2,\dots,a_{n-1} \rangle$ is a free product of cyclics of the obvious orders.

Next consider the case where p of the x_i are elements of G_1 and n-p-2 of the x_i are elements of G_2. (The remaining case where p-2 of the x_i are in G_1 and n-p of the x_i are in G_2 is handled identically.) Assume then that $x_i = a_i$, $i = 1,\ldots,n-2$. Then $G_1 = < a_1,\ldots,a_p >$ and we have $V = gWh$ where

$$g,h \in < a_{p+1},\ldots,a_{n-2} >$$

and the normal form of W (with respect to the free product of cyclics G_2) begins with a non-trivial power of a_{n-1} or of a_n and also ends with a non-trivial power of a_{n-1} or a_n since UV involves all the generators. The system $\{V,a_{p+1},\ldots,a_{n-2}\}$ is Nielsen equivalent to $\{W,a_{p+1},\ldots,a_{n-2}\}$. A product $u_1v_1 \ldots u_rv_r$ with $r \geq 1$, $1 \neq u_i \in G_1$ and

$$1 \neq v_i \in < a_{p+1},\ldots,a_{n-2} >$$

for $i = 1,\ldots,r$ can be trivial only if a product $v^{s_1}w_1 \ldots v^{s_k}w_k$ with $k \geq 1$,

$$1 \neq w_j \in < a_{p+1},\ldots,a_{n-2} >$$

and $s_k \in \mathbf{Z}/\{0\}$ for $j = 1,\ldots,k$ is trivial. But this latter product is trivial only if a product $W^{t_1}z_1 \ldots W^{t_m}z_m$ with $m \geq 1$,

$$1 \neq z_j \in < a_{p+1},\ldots,a_{n-2} >$$

and $t_j \in \mathbf{Z}/\{0\}$ for $j = 1,\ldots,m$ is trivial. But this last product can't be trivial. Therefore $u_1v_1 \ldots u_rv_r$ is non-trivial if $r \geq 1$ and $1 \neq u_i \in G_1$ and

$$1 \neq v_i \in < a_{p+1},\ldots,a_{n-2} >$$

for $i = 1,\ldots,r$. This implies that $< a_1,\ldots,a_{n-2} >$ is a free product of cyclics of the obvious orders.

We note that in general a subset of (n-1) of the given generators need not be a free product of cyclics. For example in

$$< a_1,a_2,a_3,a_4 \, ; \; a_1^{e_1} = a_2^{e_2} = a_3^{e_3} = a_4^{e_4} = a_1a_2a_3a_4 = 1 >$$

with $e_i \geq 2$ the subgroup generated by $\{a_1,a_2,a_3\}$ has the presentation

$$< a_1,a_2,a_3 \, ; \; a_1^{e_1} = a_2^{e_2} = a_3^{e_3} = (a_1a_2a_3)^{e_4} = 1 >$$

which is not a free product of cyclics. However if we allow that the relator UV involves all the generators and that U and V are proper powers in the respective free products of generators which they involve then any n-1 of the given generators generates a free product of cyclics of the obvious orders. Specifically:

Theorem 9. *Let* G *be a group of* F-*type with relator* UV *such that* UV *involves all the generators. If both* U *and* V *are proper powers* $\{U = U_1^m, V = V_1^k$ *with* $m \geq 2, k \geq 2\}$ *in the respective free products on the generators which they involve then any subset of* (n-1) *of the given generators generates a free product of cyclics of the obvious orders.*

In an orientable surface group of genus $g \geq 2$ any subgroup with three or fewer generators is a free group. An algebraic proof of this for two generators was given by G Baumslag [3] and for the case of three generators by G Rosenberger [22]. Our next result says that under certain restrictions any two generator subgroup of a group of F-type must be a free product of cyclics.

Theorem 10. *Let* G *be a group of* F-*type. Suppose further that neither* U *nor* V *is a proper power or is conjugate to a word of the form* XY *for elements* X, Y *of order* 2. *Then any two generator subgroup of* G *is a free product of cyclics.*

Proof. Without loss of generality we may assume that UV involves all the generators. Since G decomposes as a free product with amalgamation with amalgamated subgroup $A = \langle U \rangle = \langle V \rangle$ the result will follow from Lemma A if we can show that there are no non-trivial powers U^{s_1}, V^{t_1} and elements

$$X_1 \in G_1/\langle U \rangle,\ X_2 \in G_2/\langle V \rangle$$

such that

$$X_1 U^{s_1} X_1^{-1} = U^{s_2} \text{ for some } s_2 \neq 0 \text{ or } X_2 V^{t_1} X_2^{-1} = V^{t_2} \text{ for some } t_2 \neq 0.$$

Suppose there exists a non-trivial power U^s of U and an element X in $G_1/\langle U \rangle$ such that $XU^sX^{-1} = U^t$ for some $t \neq 0$. (The argument for V works identically.) We consider G_1 a discrete subgroup of $PSL_2(\mathbf{R})$ which has no parabolic elements. Then UV is hyperbolic because UV involves all generators. Let z_1, z_2 $\{z_1 \neq z_2\}$ be the fixed points of U. Then either X fixes both z_1 and z_2 or X interchanges z_1 and z_2.

If X fixes both z_1 and z_2 then X and U commute and s = t. Since G_1 is discrete this implies that $X = U_1^q$, $U = U_1^r$ for some $q,r \in \mathbf{Z}/\{0\}$ and $U_1 \in G_1$. In this case $|r| \geq 2$ since $X \in \langle U \rangle$. But this implies that U is a proper power, contrary to our assumptions.

If X interchanges z_1 and z_2 then X fixes the midpoint of the axis of U. This implies that s = -t and X has order 2. In particular $XUX^{-1} = U^{-1}$ which implies that $(XU)^2 = 1$ since $X^2 = 1$. Then $\langle X,U\rangle$ is infinite dihedral. Let Y = XU. Y has order 2 and U = XY contrary to our assumptions. Therefore there is no $X \in G_1/\langle U\rangle$ and non-trivial power U^s such that $XU^sX^{-1} = U^t$ for some $t \neq 0$. The identical argument works for V and therefore by Lemma A the theorem holds.

We note that without the additional hypotheses the theorem does not hold. For example suppose

$$G = \langle a_1,a_2,a_3,a_4 \; ; \; a_1^{e_1} = a_2^{e_2} = a_3^{e_3} = a_4^{e_4} = (a_1a_2)^8(a_3a_4)^8 = 1 \rangle.$$

Then $\langle (a_1a_2)^2,(a_3a_4)^2 \rangle$ is not a free product of cyclics. A less trivial example is given by

$$G = \langle a_1,a_2,a_3,a_4 \; ; \; a_1^2 = a_2^2 = a_3^2 = a_4^3 = a_1a_2a_3a_4 = 1 \rangle.$$

Since $[a_1a_2,a_3a_1] = a_4$ it follows that $G = \langle a_1a_2, a_1a_3 \rangle$ and thus $\langle a_1a_2,a_1a_3 \rangle$ is not a free product of cyclics.

Work of Rosenberger [24] can be used to give a complete classification in certain groups of F-type of subgroups of rank less than or equal to 4 satisfying a quadratic condition. Specifically we have:

Theorem 11. *Let* G *be a group of* F-*type such that neither* U *nor* V *is a proper power or is conjugate to a word of the form* XY *with* X, Y *elements of order* 2. *Let*

$$u_1,\ldots,u_n, \; 1 \leq n \leq 4,$$

be elements of G. *Suppose that the system* $\{u_1,\ldots,u_n\}$ *is not Nielsen equivalent to a system containing an element which is conjugate in* G *to an element of the amalgamated subgroup* $\langle U\rangle = \langle V\rangle$. *Suppose*

$$W(u_1,\ldots,u_n) = 1$$

where $W(x_1,\ldots,x_n)$ *is a quadratic word in the free group on* $x_1,\ldots,x_n$ *and let* H *be the subgroup generated by* $u_1,\ldots,u_n$. *Then* H *is either trivial, a free product of cyclics or a surface group (oriented or non-oriented).*

Finally as a direct result of the free product with amalgamation decomposition coupled with the fact that the factors are free products of cyclics we get the following result.

Theorem 12. *Let* G *be a group of* F-*type. Then any abelian subgroup is either cyclic or free abelian of rank* 2. (See Theorem 4.5 of [16].)

5. Euler characteristic for groups of F-type

One of the most powerful techniques in the study of Fuchsian groups is the Riemann-Hurwitz formula relating the Euler characteristic of the whole group to that of a subgroup of finite index. Using results of C T C Wall [29] and K Brown [6] we can define an Euler characteristic for groups of F-type and give a Riemann-Hurwitz type formula.

Theorem 13. *Let* G *be a group of* F-*type. Then* G *has a rational Euler characteristic* $\chi(G)$ *given by*

$$\chi(G) = 2 + \sum_{i=1}^{n} \alpha_i$$

where $\alpha_i = -1$ *if* $e_i = 0$ *and* $\alpha_i = -1 + (1/e_i)$ *if* $e_i \geq 2$. *If* $|G:H| < \infty$ *then* $\chi(H)$ *is defined and*

$$\chi(H) = |G:H|\chi(G) \text{ (Riemann-Hurwitz formula).}$$

Proof. In [29] C T C Wall defined a rational Euler characteristic for a class of groups which includes the trivial group, the infinite cyclic group **Z**, all finite groups and is closed under free products with finitely many factors and subgroups of finite index. If G is a group we denote this Euler characteristic by $\chi(G)$. From Wall we have that $\chi(H) = 1/|H|$ for a finite group and $\chi(H) = 0$ if H is infinite cyclic. Further if H is a free product, $H = A*B$, then

$$\chi(H) = \chi(A) + \chi(B) - 1.$$

It follows that if H is a free group of finite rank r then $\chi(H) = 1-r$. Further Wall shows that if H_1 is a subgroup of finite index in H then

$$\chi(H_1) = |H:H_1|\chi(H).$$

Now let G be a group of F-type. From Theorem 2 G is virtually torsion-free. Using this we can apply the techniques of Brown [6] to get an Euler characteristic for G by the formula

$$\chi(G) = \chi(G_1) + \chi(G_2) - \chi(A)$$

where G_1, G_2 are the factors of G from (4) and A = <U> = <V> is the amalgamated subgroup. Since A is infinite cyclic we have $\chi(A) = 0$ and therefore $\chi(G) = \chi(G_1) + \chi(G_2)$. Further G_1 and G_2 are free products of cyclic groups so we can apply the computation rules of Wall above to obtain

$$\chi(G) = 2 + \sum_{i=1}^{n} \alpha_i$$

where $\alpha_i = -1$ if $e_i = 0$ and $\alpha_i = -1 + (1/e_i)$ if $e_i \geq 2$. We can also write this as

$$\chi(G) = 2 - n + \sum_{i=1}^{n} \beta_i$$

where $\beta_i = 0$ if $e_i = 0$ and $\beta_i = 1/e_i$ if $e_i \geq 2$. This second form is somewhat closer to the form usually written for F-groups. The Riemann-Hurwitz formula follows directly from Wall.

We note further that the Euler characteristic of a group of F-type can be zero. In fact $\chi(G) = 0$ if G is a 2-generator group of F-type.

6. The generalized word problem for groups of F-type

F Tang in [28] has proven the following general result on one-relator products of cyclics whose relators have the form $(UV)^t$ with $t \geq 1$ and U, V cyclically reduced words involving disjoint subsets of the generators. Such relators he calls *cyclically pinched.*

Theorem 14 (Tang [28]). *Let*

$$G = < x_1,\ldots,x_n \,;\, x_i^{e_i} = 1,\ i = 1,\ldots,n,\ (UV)^t = 1 >$$

with $e_i = 0$ *or* $e_i \geq 2$ *for* $i = 1,\ldots,n$,

$$U = U(a_1,\ldots,a_p),\ V = V(a_{p+1},\ldots,a_n)$$

non-trivial cyclically reduced words in the free products of generators they involve and $1 < p < n$ *and* $t \geq 1$. *Then* G *has a solvable generalized word problem.*

In particular using t = 1 we obtain:

Corollary. *Let* G *be a group of* F-*type. Then* G *has a solvable generalized word problem.*

In [7] Comerford and Truffault make the observation that Fuchsian groups have solvable conjugacy problem. Whether this is true for groups of F-type is at present unknown and we leave it as a question.

7. Closing questions

In the final section we give a selection of open problems related to results in Fuchsian groups about groups of F-type and groups similar to groups of F-type.

(1) Are groups of F-type conjugacy separable?

In recent work Fine and Rosenberger [12] using some results of Allenby and Tang [1] and Stebe [27] proved that all Fuchsian groups are conjugacy separable.

(2) Suppose that either U or V is a proper power. Describe additional conditions if possible on the resulting group G so that there exists a faithful representation in $PSL_2(C)$.

(3) Suppose neither U nor V is a proper power. Describe conditions on the group G such that G has a faithful representation $\rho: G \to PSL_2(C)$ with $\rho(G)$ discrete.

(4) Suppose either U or V is a proper power. Describe additional conditions if any on the resulting group G so that G is residually finite or Hopfian.

(5) Let G be a group of F-type. Generalize Theorem 7 and give a complete description of the two-generator subgroups of G.

(6) Let G be a group of F-type. Describe the solvable subgroups of G.

(7) Let G be a group of F-type. Describe additional conditions if any on UV such that one or more of the following holds:

(a) Subgroups of finite index have the same presentation "form" as G.

(b) Torsion-free subgroups of finite index are one-relator groups.

(c) Subgroups of infinite index are free products of cyclics.

Along the lines of part (b) we make the following conjecture.

Conjecture. *Let* G *be a group of* F-*type and suppose* UV *involves all the generators. Assume that each torsion-free subgroup of finite index has a one-relator presentation. Then* G *is a planar discontinuous group.*

(8) Let G be a group of F-type. Is the conjugacy problem for G solvable?

(9) Let G be a group of F-type. Suppose rg(G) is the algebraic rank of G. Is it true that $n-2 \leq rg(G) \leq n$?

(10) Let G be a group of F-type. Is the automorphism group Aut(G) finitely generated or finitely presented?

(11) Let G be a group of F-type. Under what conditions is each automorphism induced by an automorphism of the free group of rank r(G)?

This is true in the following cases:

(i) G is a torsion-free cyclically pinched one-relator group - that is all $e_i = 0$.

(ii) G is a cyclically pinched one-relator group with torsion which is not a free product of cyclics - that is at most one $e_i \geq 2$.

(iii) The relator UV has the form

$$UV = a_1 \ldots a_m(a_{m+1}^{j_{m+1}} \ldots a_k^{j_k}[a_{k+1}, a_{k+2}] \ldots [a_{n-1},a_n])^t$$

where $2 \leq m \leq k \leq n$, $t \geq 1$, $e_j \geq 2$ for $i = 1,\ldots,m$, $e_i = 0$ for $i = m+1,\ldots,n$, $j_i \geq 2$ for $i = m+1,\ldots,k$ and $m \geq 3$. If $m = n$ then

$$((1/e_1) + (1/e_2) + (1/e_3)) \leq 1.$$

In particular this holds if G is a Fuchsian group. Further if at most one $e_i \geq 2$ then G has only finitely many Nielsen inequivalent one-relator presentations and the isomorphism problem for G is solvable. This means that we may decide in finitely many steps whether a one-relator presentation is isomorphic to G or not. (See [22] and [23].)

(12) What is the virtual cohomological dimension of a group of F-type? It was pointed out to us by W Bogley and D Collins that if G is a group of F-type then its virtual cohomological dimension should be ≤ 2.

References

1. R B J T Allenby & C Y Tang, Conjugacy separability of certain 1-relator groups with torsion, *J. Algebra* **103** (1986), 619-637.

2. B Baumslag & S J Pride, Groups with two more generators than relators, *J. London Math. Soc.* (2) **17** (1987), 425-426.

3. G Baumslag, On generalized free products, *Math. Z.* **78** (1962), 423-438.

4. G Baumslag, J Morgan & P Shalen, Generalized triangle groups, *Math. Proc. Cambridge Philos. Soc.* **102** (1987), 25-31.

5. A F Beardon, *The Geometry of Discrete Groups* (Springer-Verlag, 1983).

6. K S Brown, Groups of virtually finite dimension, in *Homological Group Theory* (Proceedings of the 1977 Durham Symposium, London Math. Soc. Lecture Notes Series **36**, 1979), 27-70.

7. L Comerford & B Truffault, The conjugacy problem for free products of sixth groups with cyclic amalgamation, *Math. Z.* **149** (1976), 169-181.

8. B Fine, J Howie & G Rosenberger, One-relator quotients and free products of cyclics, *Proc. Amer. Math. Soc.* **102** (1988), 1-6.

9. B Fine & G Rosenberger, Complex representations and one-relator products of cyclics, in Geometry of Group Representations, *Contemp. Math.* **74** (1987), 131-149.

10. B Fine, F Levin & G Rosenberger, Free subgroups and decompositions of one-relator products of cyclics; Part 1 : The Tits alternative, *Arch. Math.* **50** (1988), 97-109.

11. B Fine, F Levin & G Rosenberger, Free subgroups and decompositions of one-relator products of cyclics; Part 2 : Normal torsion-free subgroups and FPA decompositions, *J. Indian Math. Soc.* **49** (1985), 237-247.

12. B Fine & G Rosenberger, Conjugacy separability of Fuchsian groups, *Contemp. Math.*, to appear.

13. G Higman, H Neumann & B H Neumann, Embedding theorems for groups, *J. London Math. Soc.* **24** (1949), 247-254.

14. R C Lyndon & P E Schupp, *Combinatorial Group Theory* (Springer-Verlag. 1977).

15. W Magnus, *Non-Euclidean Tesselations and their Groups* (Academic Press, 1974).

16. W Magnus, A Karass & D Solitar, *Combinatorial Group Theory* (Wiley Interscience, 1968).

17. A I Malcev, On faithful representations of infinite groups of matrices, *Amer. Math. Soc. Trans.* (2) **45** (1965), 1-18.

18. P M Neumann, The SQ-universality of some finitely presented groups, *Proc. Camb. Phil. Soc.* **16** (1973), 1-6.

19. J Ratcliffe, Euler characteristics of 3-manifold groups and discrete subgroups of SL(2,C), preprint.

20. G Rosenberger, Faithful linear representations and residual finiteness of certain one-relator products of cyclics, *J. Siberian Math. Soc.*, to appear.

21. G Rosenberger, The SQ-universality of one-relator products of cyclics (Proc. of the Int. Conf. on Topology, Baku, 1987), to appear.

22. G Rosenberger, *Zum Rang und Isomorphieproblem fur freie Produkte mit Amalgam* (Habilitationsschrift, Hamburg 1974).

23. G Rosenberger, On one-relator groups that are free products of two free groups with cyclic amalgamation, in *Groups St Andrews 1981* (Cambridge University Press, 1983), 328-344.

24. G Rosenberger, Bemerkungen zu einer Arbeit von R C Lyndon, *Archiv. der Math.* **40** (1983), 200-207.

25. A Selberg, On discontinuous groups in higher dimensional symmetric spaces (Int. Colloq. Function Theory, Tata Institute, Bombay 1960), 147-164.

26. P Shalen, Linear representations of certain amalgamated products, *J. Pure Appl. Algebra* **15** (1979), 187-197.

27. P Stebe, Conjugacy separability of certain Fuchsian groups, *Trans. Amer. Math. Soc.* **163** (1972), 173-188.

28. C Y Tang, Some results on one-relator quotients of free products, *Contemp. Math.*, to appear.

29. C T C Wall, Rational Euler characteristics, *Proc. Camb. Phil. Soc.* **57** (1961), 182-188.

30. H Zieschang, Uber die Nielsensche Kurzungsmethode in freien Produkten mit Amalgam, *Invent. Math.* **10** (1970), 4-37.

A THEOREM ON FREE PRODUCTS OF SPECIAL ABELIAN GROUPS

ANTHONY M GAGLIONE

U S Naval Academy, Annapolis, MD 21402, USA

HERMANN V WALDINGER

Polytechnic University, Brooklyn, NY 11201, USA

1. Introduction

In a previous paper [1], the quotient groups of the lower central series $\bar{G}_n = G_n/G_{n+1}$ were studied. There the group G was assumed to be a free product of a finite number of finitely generated Abelian groups and G_n denoted the nth subgroup of the lower central series of G. Here we give an improved proof of a complicated lemma which first appeared in [1] (in particular, Lemma 4.4 of [1]). The proof given here, especially for property (iii) of the conclusion of that lemma, is a significant simplification of that which appears in [1]. We observe that one of the consequences of Lemma 4.4 of [1] is to give a set of free generators for the lower central quotients in the case where the free factors are torsion free (i.e., G = J in the terminology of [1]. Moreover the free generators are the J-basic commutators also using the terminology of Definition 4.1 of [1]). The authors only thought of this simplification after the publication of [1]. Furthermore, our improved proof uses results from [2] and [3].

In this paper, we will employ the notation, terminology, results, references, and equations of [1]. Furthermore, the numbering, a.b, of any definition, equation, etc., of [1] will correspond here to a.b-I. For example, Lemma 2.1-I will mean Lemma 2.1 of [1].

2. Preliminaries

In order to carry out our goal, we need to give some preliminary machinery.

Lemma 2.1. (cf. Corollary 5.1 [3]). *Let* S *be the subgroup of* F *generated by a finite set* Σ *of* F*-simple commutators (see Definition 3.6-I) of weight* > 1. *Let* $U = \{u_1,u_2,\ldots,u_m\} \subseteq \Sigma$. *Let* $V = \{v_1,v_2,\ldots,v_m\}$ *be a set of* m *nontrivial elements*

v_i *of* F *such that* $W(v_i) \geq W(u_i)$ *for* $1 \leq i \leq m$. *(Here we are using the notation of Definition 2.1-I for weight.) Let* Ψ *be the mapping of* Σ *given by*

$$u_i \to v_i \text{ if } u_i \in U$$

$$c_i \to c_i \text{ if } c_i \notin U.$$

Then Ψ *extends to a homomorphism, also denoted by* Ψ, *of* S *into* F *such that if* $u \in S$ *and* $W(u) = \omega > 0$, *then its image* $v = \Psi(u) \in F_\omega$.

We will also require:

Lemma 2.2. (cf. Theorem 1.1[2]) *Let* c *be the* F-*simple basic commutator of weight* $n > 1$,

$$c = (c_{j_1}, c_{j_2}, \ldots, c_{j_n}). \tag{2.1}$$

Let d *be a generator of* F *such that* $d < c_{j_n}$. *Let* $v = (c_{j_2}, d, e_1, e_2, \ldots, e_{n-1})$ *be such that* $e_1 = c_{j_1}$ *for* $n = 2$, *but* $e_1 \leq e_2 \leq \ldots \leq e_{n-1}$ *is a rearrangement of* $c_{j_1}, c_{j_3}, \ldots, c_{j_n}$, *for* $n > 2$. *(Note that* v *is* F-*simple of weight* n+1 *if* $d < c_{j_2}$.*) Then the following identity holds in* F

$$(c,d) = \begin{cases} \Pi_1 \, M(c,d) \, \Pi_2 & \text{if } d \geq c_{j_2} \text{ (Case I)} \\ \Pi_1 \, v^{-1} \, \Pi_2 \, M(c,d) \, \Pi_3 & \text{if } d < c_{j_2} \text{ (Case II)} \end{cases} \tag{2.2}$$

where the Π_j $(j = 1,2,3)$ *are either* = 1 *or are words in finitely many* F-*simple commutators* u_i *such that*

(i) $1 < W(u_i) < n + 1$;

(ii) *if* $u_i = (v_1, v_2, \ldots, v_t)$, *then* $v_1, v_2, \ldots, v_t$ *is a rearrangement of a subsequence of* $c_{j_1}, \ldots, c_{j_n}, d$.

3. Statement and proof of the main result

We are now ready to state and prove our main result:

Theorem 3.1. (See Lemma 4.4-I.) *Let* c *be* F-*simple as in* (2.1). *Suppose that* c *satisfies Criterion* 1 *but not Criterion* 3. *(See Definition 4.1-I). Let* t *be the largest integer such that* c_{j_t} *occurs in* c *and* $(c_{j_t}, c_{j_1}) = 1$ *in* J. *(The hypothesis that* c *does not satisfy Criterion* 3 *guarantees that such a* t *exists,* $t \geq 3$, *and that* $j_t > j_1$.*) Let the* F-*simple commutator* d(c) *be defined by*

$$d(c) = (c_{j_t}, d_2, d_3, \ldots, d_n) \tag{3.1}$$

where $d_2, d_3, \ldots, d_n$ *is that rearrangement of* $c_{j_1}, \ldots, c_{j_{t-1}}, c_{j_{t+1}}, \ldots, c_{j_n}$ *for which* $d_2 \leq \ldots \leq d_n$. *Then the following holds in* J

$$c = H_1\, d(c)\, H_2 \tag{3.2}$$

where the H_j $(j = 1,2)$ *are words in J-simple commutators* u_i *such that*

(i) $1 < W(u_i) < n$;

(ii) *if* $u_i = (v_1, \ldots, v_k)$, *then* $v_1, \ldots, v_k$ *is a rearrangement of a subsequence of* $c_{j_1}, \ldots, c_{j_n}$;

(iii) $W(H_1\, d(c)\, H_2) \geq n$ *in* F.

Proof. We proceed by induction on the weight of c, W(c). Clearly, $W(c) = n \geq 3$. For $n = 3$, we have $c = (c_{j_1}, c_{j_2}, c_{j_3})$ with $(c_{j_1}, c_{j_3}) = 1$ in J and $c_{j_3} > c_{j_1}$. Thus equation (3.20)-I yields that in J

$$c = (c_{j_1}, c_{j_2})^{-1}(c_{j_3}, c_{j_2})(c_{j_3}, c_{j_2}, c_{j_1})\,(c_{j_1}, c_{j_2})(c_{j_3}, c_{j_2})^{-1}. \tag{3.3}$$

This shows that we have (3.2) with properties (i) and (ii) for $n = 3$. In order to show that the right hand side of (3.3) also satisfies property (iii), we let

$$a = (c_{j_1}, c_{j_2}),\ b = (c_{j_3}, c_{j_2}),\ \text{and } A = (c_{j_3}, c_{j_2}, c_{j_1}).$$

Then (3.3) becomes

$$c = a^{-1} b\, A\, a\, b^{-1}$$

and we need to show that $W(a^{-1} b\, A\, a\, b^{-1}) \geq 3$ in F. But in F

$$a^{-1}\, b\, A\, a\, b^{-1} = (a^{-1}\, b\, A\, b^{-1} a)\cdot(a, b^{-1}) = R\cdot(a, b^{-1})$$

where $R = a^{-1}\, b\, A\, b^{-1}\, a$. Evidently by (2.3a)-I, $W(a, b^{-1}) \geq 4$ in F so that

$$W(a^{-1}\, b\, A\, a\, b^{-1}) = W(R\cdot(a, b^{-1})) \geq W(R) \geq 3$$

in F. This proves the result for $n = W(c) = 3$.

Next, suppose that we have already proven the result for all c with $3 \leq W(c) < k$ which satisfy the hypothesis of our theorem. We then find the smallest integer $n \geq k$ such that there exist commutators, c, which satisfy the hypothesis of our theorem and have $W(c) = n$. We will prove our theorem for all such c.

We now define $\delta = n - t$ and we treat two cases: $\delta > 0$ and $\delta = 0$. For $\delta > 0$, let us write $e = c^L$ and we note that, in J, $e = (c_{j_1}, \ldots, c_{j_{n-1}})$ is by the induction hypothesis a word of the form (3.2), i.e.,

$$e = H_{11}\, d(e)\, H_{12},$$

where H_{1j} $(j = 1,2)$ are words in J-simple basic commutators z_i satisfying (i) and (ii) of our conclusion, and $W(H_{11}\, d(e)\, H_{12}) \geq n-1$ in F. In particular every z_i

satisfies $1 < W(z_i) < n-1$ and is such that $z_i^R \le c_{j_{n-1}}$ (recall by the hypothesis of our theorem that $c_{j_1} < c_{j_t}$ and since $\delta > 0$ that $c_{j_t} \le c_{j_{n-1}}$). Applying the trivial identity (3.19)-I to the computation of $c = (e, c_{j_n}) = (H_{11}\ d(e)\ H_{12}, c_{j_n})$, we find that in J

$$c = \Pi_1\ (d(e), c_{j_n})\ \Pi_2 \tag{3.4}$$

where the Π_j ($j = 1,2$) are words in z_i, (z_i, c_{j_n}), and $d(e)$. (Note $d(e)$ only occurs in Π_1.) Now $d(c) = (d(e),c_{j_n})$ by (3.1) since $\delta > 0$. Here we denote the word on the right hand side of (3.4) by w and note that w is a word in the F-simple basic commutators $d(e)$, z_i, (z_i, c_{j_n}), and $d(c)$, i.e.,

$$w = \Pi_1\ d(c)\ \Pi_2 \tag{3.4a}$$

where the Π_j ($j = 1,2$) are as in (3.4). Moreoever, we note that $W(z_i, c_{j_n}) \le n-1$. So that either the F-simple commutators (z_i, c_{j_n}) are all J-simple or we may use the induction hypothesis to write those (z_i, c_{j_n}) which are not J-simple as words of the form (3.2) in J-simple commutators. After making the replacements for those (z_i, c_{j_n}) in (3.4a) which are not J-simple, we find that in J

$$c = H_1\ d(c)\ H_2 \tag{3.5}$$

where we have demonstrated that the H_j ($j = 1,2$) are words in J-simple commutators satisfying conditions (i) and (ii) of our conclusion.

To see that our expression for c in (3.5) also satisfies condition (iii) of our conclusion, we first note that the word w as given in (3.4a) has $W(w) \ge n$ in F. (This is true since $w = (H_{11}\ d(e)\ H_{12}, c_{j_n})$ in F and $W(H_{11}\ d(e)\ H_{12}) \ge n-1$.) Thus if all the (z_i, c_{j_n}) satisfy Criterion 3, i.e., are J-simple, we are done. Otherwise let Σ be set of all the distinct F-simple basic commutators which occur in w as given by (3.4a). Let

$$U = \{u_1, \ldots, u_m) \subseteq \Sigma$$

be those (z_i, c_{j_n}) in (3.4a) which are not J-simple and let $\{v_1, \ldots, v_m\}$ be those words in J-simple commutators which are obtained by applying the induction hypothesis to the (z_i, c_{j_n}). Thus $u_i = v_i$ in J and

$$W(u_i) \le W(v_i) \text{ in F for all } i = 1, \ldots, m.$$

Now to go from (3.4) to (3.5), we replace the u_i by the v_i in w as given by (3.4a), i.e., we apply the map $\Psi : u_i \to v_i$. Thus since $w \in F_n$, Lemma 2.1 implies that

$$\Psi(w) = H_1\, d(c)\, H_2$$

in (3.5) is such that $(H_1\, d(c)\, H_2) \in F_n$, i.e., $W(H_1\, d(c)\, H_2) \geq n$. This completes the proof of our result in the case $\delta > 0$.

It remains to consider the case $\delta = 0$ for which we write

$$(c^L)^L = A,\ (c^L)^R = c_{j_{n-1}} = g, \text{ and } c^R = c_{j_n} = h.$$

Then identity (3.20)-I gives

$$c = (A,g,h) = (A,g)^{-1}(A,h)^{-1}A^{-1}(h,g)A(A,g)(A,h)(A,h,g)(h,g)^{-1}. \tag{3.6}$$

Either $(h,g) = 1$ in J or it is J-simple. Also all the commutators A, (A,g), and (A,h) are F-simple of weight $< n$. Hence any such commutator is either J-simple or can be written in J by the induction hypothesis as a word in J-simple commutators satisfying conditions (i), (ii) and (iii) of our conclusion. Letting $Z = (A, h, g)$ in (3.6), we claim that it is sufficient to prove that Z can be written in J as a word in J-simple commutators in the form

$$Z = H_{11}\, d(c)\, H_{12} \tag{3.7}$$

where the H_{1j} $(j = 1,2)$ are words in J-simple commutators, u_i, satisfying (i) and (ii) of our conclusion, and

$$W(H_{11}\, d(c)\, H_{12}) \geq n.$$

To see this, suppose we substitute (3.7) and the expressions in terms of J-simple commutators (for those among A, (A,g), and (A,h) which are not J-simple) into (3.6). This writes c in J as

$$c = H_1\, d(c)\, H_2 \tag{3.8}$$

where the H_j $(j = 1,2)$ are words in J-simple commutators. Clearly, conditions (i) and (ii) of our conclusion are satisfied by (3.8), according to the induction hypothesis and our claim.

For condition (iii), we note that (3.6) is an identity in F. Thus the right hand side of (3.6) has weight n in F since c is F-simple of weight n. Now let $(h,g) = d$, $(A,g) = a$, and $(A,h) = b$. Then (3.6) becomes

$$c \;= a^{-1}\, b^{-1}\, A^{-1}\, d\, A\, a\, b\, Z\, d^{-1}$$

$$= (a,b)\,(A\, a\, b,\, d^{-1}) \cdot d\, Z\, d^{-1}. \tag{3.6a}$$

By (2.3a)-I, $a \in F_{n-1}$, $b \in F_{n-1}$, $A \in F_{n-2}$, $Z \in F_n$, and $d \in F_2$. (Either $d = 1$ or $W(d) = 2$.) Let $\tilde{a}, \tilde{b}, \tilde{A}, \tilde{Z}, \tilde{d}$ be the words in J-simple commutators such that

$$a = \tilde{a},\ b = \tilde{b},\ A = \tilde{A},\ Z = \tilde{Z},\ d = \tilde{d}$$

in J (i.e., $\tilde{Z} = H_{11}\, d(c)\, H_{12}$) as in (3.7)). We know by the induction hypothesis and our claim for Z that $\tilde{a} \in F_{n-1}$, $\tilde{b} \in F_{n-1}$, $\tilde{A} \in F_{n-2}$, $\tilde{Z} \in F_n$. Also $\tilde{d} \in F_2$ since d either is 1 in J or it is a J-simple commutator of weight 2. (Here $\tilde{d} = d$ if $d \neq 1$ in J but $\tilde{d} = 1$ if $d = 1$ in J.) Letting

$$\tilde{c} = (\tilde{a}, \tilde{b})\, (\tilde{A}\, \tilde{a}\, \tilde{b}, \tilde{d}^{-1}) \cdot \tilde{d}\ \tilde{Z} \tilde{d}^{-1},$$

we note that $(\tilde{a}, \tilde{b}) \in F_{2n-2} \subseteq F_n$ (since $2n-2 > n$ in our case $n > 3$),

$$(\tilde{A}\, \tilde{a}\, \tilde{b}, \tilde{d}^{-1}) \in F_n$$

according to (2.3a)-I, and $\tilde{d}\ \tilde{Z} \tilde{d}^{-1} \in F_n$. Thus $W(\tilde{c}) \geq n$ in F. Since

$$\tilde{c} = H_1\, d(c)\, H_2 \text{ in } F,$$

evidently our conclusion that $W(H_1\, d(c)\, H_2) \geq n$ in F holds.

Thus we now focus our attention on $Z = (A,h,g)$, and prove the claim for Z as given by (3.7). As already mentioned, $b = (A,h)$ is, by the induction hypothesis, a word, $\tilde{b}$, in J of the form (3.2) in d(b) and J-simple commutators, s_i, of weight less than n-1 with all the special properties of our conclusion. Therefore $Z = (b,g)$ can be written by the trivial identity (3.19)-I in J in the form

$$Z = (\tilde{b}, g) = H_{01}(d(b), g)H_{02} \tag{3.9}$$

where the H_{0j} $(j = 1,2)$ are words in d(b), s_i, and (s_i, g). By the induction hypothesis $W(\tilde{b}) \geq n-1$, hence (2.3a)-I implies that the right hand side of (3.9), $H_{01}\, (d(b), g)H_{02}$, has weight at least n in F.

Let us now examine (d(b), g). From (3.1) and the fact that $\delta = 0$,

$$d(b) = (c_{j_n}, e_2, e_3, \ldots, e_{n-1})$$

where $e_2 \leq e_3 \leq \ldots \leq e_{n-1}$ is a rearrangement of $c_{j_1}, c_{j_2}, \ldots, c_{j_{n-2}}$. (Evidently $c_{j_2} = e_2$, since $c_{j_1} > c_{j_2} \leq c_{j_3} \leq \ldots \leq c_{j_n}$.) We note from Lemma 3.5-I that $M(d(b),g) = d(c)$. Next, we apply Lemma 2.2 to write (d(b),g) as $\Pi_1\, d(c)\, \Pi_2$ in F. (Note that $g = c_{j_{n-1}} \geq e_2$ so that we are in Case I of Lemma 2.2 and also that if $g \geq e_{n-1}$, then $(d(b),g) = d(c)$ and we just then take $\Pi_1 = \Pi_2 = 1$.)

Let us also apply Lemma 2.2 to rewrite those (s_i, g) in (3.9) which are not basic (i.e., $g < s_i^R$ so that Lemma 2.2 applies) in terms of F-simple commutators. Thus

$$(s_i, g) = \Pi_{i1}\, M(s_i, g)\, \Pi_{i2}$$

in F. (Here it does not matter which case of (2.2) we are in because $W(s_i, g) < n$.) Thus Π_1, Π_2, and Π_{ij} ($j = 1,2$) are words in F-simple commutators, z_i, all of which have $1 < W(z_i) < n$ and satisfy condition (ii) of our conclusion. (We also now allow z_i to be any of the $M(s_i,g)$ where (s_i,g) is not basic.)

Substituting these into (3.9), we have that Z can be written in J as a word, w, in F-simple commutators d(c), s_i, (s_i,g) (only those which are basic are included here), and z_i, i.e., in J we have

$$Z = w = \Omega_{01}\, d(c)\, \Omega_{02} \tag{3.10}$$

where Ω_{0j} ($j = 1,2$) are words in the s_i, (s_i, g) - only the basic ones - and z_i. We may now use the induction hypothesis to write those F-simple commutators, the (s_i, g) and z_i, which occur in (3.10) and are not J-simple, as words in J-simple commutators. Making these replacements in (3.10) gives (3.7). To see this, we note that now Z is written, in J, in terms of J-simple commutators, u_i, which clearly satisfy (i) and (ii) of our conclusion.

To see that $W(H_{11}\, d(c)\, H_{12}) \geq n$ in F, we first note that the right hand side of (3.9) is equal to the right hand side of (3.10) in F, i.e.,

$$w = H_{01}\, (d(b), g)\, H_{02}$$

holds in F. Thus $w \in F_n$. We can now continue in a manner exactly analogous to that used in the case of $\delta > 0$, because the situation in (3.10) is similar to that of equation (3.4a). In this way Lemma 2.1 shows that the weight of $H_{11}\, d(c)\, H_{12}$ given in (3.7) is at least n. This completes the proof of our claim for Z given in (3.7) and with that the proof of our theorem is accomplished.

References

1. A M Gaglione & H V Waldinger, Factor groups of the lower central series of free products of finitely generated Abelian groups, in *Proc. Groups St Andrews, 1985* (Cambridge University Press, 1986), 164-203.
2. A M Gaglione & H V Waldinger, A theorem in the commutator calculus, in *Proc. Singapore Group Theory Conference 1987* (Walter de Gruyter, 1988), 367-371.
3. A M Gaglione & H V Waldinger, The commutator collection process, in *Proc. AMS Special Session on Combinatorial Group Theory, 1988* (Amer. Math. Soc. Providence), to appear.

SCHUR ALGEBRAS AND GENERAL LINEAR GROUPS

J A GREEN

Mathematics Institute, Warwick University, Coventry CV4 7AL

Introduction

In his doctoral dissertation of 1901, I Schur posed and solved completely the problem which can be expressed in modern terms as follows: *Given a positive integer* n, *to find all finite-dimensional polynomial representations of the general linear group* $GL_n(C)$. By definition a *representation* R of $GL_n(C)$ of dimension N is a homomorphism of groups

$$R : GL_n(C) \rightarrow GL_N(C),$$

and R is *polynomial* if each coefficient $r_{\alpha\beta}(g)$ of the matrix

$$R(g) = (r_{\alpha\beta}(g))_{\alpha,\beta\in N}$$

can be expressed as a polynomial, with complex coefficients, in the n^2 entries $g_{\mu\nu}$ of the element $g = (g_{\mu\nu})_{\mu,\nu\in N}$ of $GL_n(C)$. If, moreover, all these polynomials $r_{\alpha\beta}$ are homogeneous (in the n^2 "variables" $g_{\mu\nu}$) of some fixed degree r, we say R is *homogeneous of degree* r. Schur showed that every polynomial representation of $GL_n(C)$ is equivalent to a direct sum of homogeneous ones, so that nothing is lost if we confine ourselves to polynomial representations of $GL_n(C)$ which are homogeneous of some fixed degree r. Schur showed that each such representation R gives rise to a family R_0 of matrices which obey certain multiplication rules, so that R_0 can be interpreted as a matrix representation of a finite-dimensional complex linear algebra, which is now often called the *Schur algebra* $S_C(n,r)$. Conversely, each representation R_0 of $S_C(n,r)$ can be used to construct a polynomial representation R of $GL_n(C)$ of degree r; the correspondences

$$R \rightarrow R_0, \quad R_0 \rightarrow R$$

are inverse to each other, and respect the representation - theoretical properties of equivalence and irreducibility. Then, in a second application of this same principle, Schur showed that, if $n \geq r$, there exist correspondences $R_0 \leftrightarrow R_1$

connecting the representations R_0 of $S_C(n,r)$ with the representations R_1 of the symmetric group $P(r)$ on the set $\mathbf{r} = \{1,2,\ldots,r\}$. A few years earlier, Frobenius had founded the character theory of finite groups (1896), and had determined the irreducible characters of the symmetric group $P(r)$ (1900). Schur used these new discoveries with astonishing effect. By combining Frobenius's work with the correspondences

$$R \leftrightarrow R_0 \leftrightarrow R_1,$$

Schur was able to show that every polynomial representation R of $GL_n(C)$ is completely reducible, that the equivalence classes or types of irreducible polynomial representations of $GL_n(C)$ of degree r can be parametrized by the partitions $\lambda = (\lambda_1 \geq \ldots \geq \lambda_n \geq 0)$ of r into not more than n parts, and that the character of an irreducible polynomial representation R_λ of $GL_n(C)$ of type λ, expressed as function of the eigenvalues $X_1, \ldots, X_n$ of a generic matrix g of $GL_n(C)$, is the symmetric polynomial

$$s_\lambda(X_1,\ldots,X_n) = \frac{|X_i^{\lambda_j+n-j}|}{|X_i^{n-j}|}$$

which is now called the *Schur function* associated to λ.

Example $\mathbf{n = r = 2}$. Corresponding to the two partitions $\lambda = (2,0)$, $\lambda = (1,1)$ of $r = 2$, are irreducible representations

$$R_{(2,0)} : g \to \begin{bmatrix} g_{11}^2 & g_{11}g_{12} & g_{12}^2 \\ 2g_{11}g_{21} & g_{11}g_{22} + g_{12}g_{21} & 2g_{12}g_{22} \\ g_{21}^2 & g_{21}g_{22} & g_{22}^2 \end{bmatrix},$$

$$R_{(1,1)} : g \to (\det g) = (g_{11}g_{22} - g_{12}g_{21}),$$

of dimensions 3, 1 respectively. To calculate the character of R_λ, the easiest way is to put $g = \begin{pmatrix} X_1 & 0 \\ 0 & X_2 \end{pmatrix}$ in $R_\lambda(g)$ and then evaluate the trace: we get

$$s_{(2,0)}(X_1,X_2) = X_1^2 + X_1X_2 + X_2^2, \quad s_{(1,1)}(X_1,X_2) = X_1X_2.$$

There have been several attempts to extend Schur's work to fields K of finite characteristic. If we assume (as we shall do throughout these lectures) that K is an *infinite* field, then the correspondence $R \leftrightarrow R_0$ between polynomial representations R of $GL_n(K)$ of degree r, and representations R_0 of the Schur algebra $S_K(n,r)$, still holds (see Ch. 3, Section (3.1)). But the correspondence

$$R_0 \leftrightarrow R_1$$

is no longer as sharp as it was for K = C, and the vital property of complete reducibility fails. The irreducible polynomial representations of $GL_n(K)$ of degree r can still, surprisingly, be labelled by the partitions of r, but at present no character formula for them is known.

In fact the greatest successes in this problem have been in finding *integral forms* of Schur's representations R_λ - these are polynomial representations of $GL_n(C)$ in which all the *coefficients* of the polynomials $r_{\alpha\beta}$ are integers, so that one can interpret them as representations of $GL_n(K)$, where K is any field, or indeed any commutative ring. For example, $R_{(2,0)}$ and $R_{(1,1)}$ just described are integral in this sense. But there may be many integral forms of the same complex representation, for example

$$R'_{(2,0)} : g \rightarrow \begin{bmatrix} g_{11}^2 & 2g_{11}g_{12} & g_{12}^2 \\ g_{11}g_{21} & g_{11}g_{22} + g_{12}g_{21} & g_{12}g_{22} \\ g_{21}^2 & 2g_{21}g_{22} & g_{22}^2 \end{bmatrix}$$

is equivalent over C to $R_{(2,0)}$, but as representations of $GL_2(K)$, where char K = 2, it turns out that $R_{(2,0)}$ and $R'_{(2,0)}$ are inequivalent (and both are reducible).

Integral forms for the R_λ, although in the language of modules rather than of matrix representations, have been described recently by a number of people, in particular by Carter and Lusztig (1974), James (1978), Clausen (1979), and Akin and Buchsbaum (1985). In these lectures is given an account of Clausen's "Weyl modules of type 1", although under Akin and Buchsbaum's name of "Schur modules". The first two chapters describe these modules and give their bases of standard bideterminants - the treatment follows Clausen (1989) and de Concini, Eisenbud and Procesi (1980). The third chapter introduces the Schur algebra $S_K(n,r)$, and gives some applications to the polynomial representation theory of $GL_n(K)$.

I. Combinatoric Methods

(1.1) Let n and r be positive integers, and let K be an infinite field of arbitrary characteristic. The aim of Chapters I and II of these lectures is to describe, for each partition

$$\lambda = (\lambda_1 \geq \ldots \geq \lambda_n \geq 0)$$

of r into not more than n (positive) parts $\lambda_1, \lambda_2, \ldots$, a K-space $L_\lambda = L_\lambda(K)$ on which $GL_n(K)$ acts - we call L_λ the *Schur module* for $GL_n(K)$ and λ. (Cf. Akin and Buchsbaum [AB], who use this term for the module which we call $L_{\tilde{\lambda}}$ ($\tilde{\lambda}$ is the conjugate of λ).) The construction of L_λ provides it with a natural K-basis, and in case K = C, this basis affords a matrix representation of $GL_n(C)$ which is an integral form of Schur's irreducible representation R_λ. Moreover the family $\{L_\lambda(K)\}$, K ranging over all infinite fields, is "defined over **Z**" in the sense of [G, p.30]; in fact $L_\lambda(K)$ is isomorphic (as $GL_n(K)$-module) to the module $D_{\lambda,K}$ of [G], but we shall construct $L_\lambda(K)$ by Clausen's combinatoric method (see [Cl, Def. 5.1(4), p.180]), which has many advantages over that used in [G].

(1.2) It is worth recalling a very general principle of representation theory: if a group G acts on a set $\mathcal{M}$ on the left, so that a "product" $gm \in \mathcal{M}$ is defined for each $g \in G$ and $m \in \mathcal{M}$, satisfying the usual rules

$$g(g'm) = (gg')m \text{ and } 1m = m,$$

then, for any field K, G also acts on the set $\mathcal{F}$ of all maps $\phi : \mathcal{M} \to K$. This new action is on the right: if $\phi \in \mathcal{F}$ and $g \in G$, define $\phi \circ g \in \mathcal{F}$ by

$$(\phi\circ g)(m) = \phi(gm), \text{ all } m \in \mathcal{M}.$$

The set $\mathcal{F}$ inherits from K the structure of a commutative K-algebra, the necessary operations on $\mathcal{F}$ being defined "pointwise", i.e. if $c,c' \in K$, $\phi,\phi' \in G$, and 1_K is the identity element of K, then $c\phi + c'\phi'$, $\phi\phi'$ and **1** are by definition the elements of $\mathcal{F}$ which take an arbitrary $g \in G$ to

$$c\phi(g) + c'\phi'(g), \quad \phi(g)\phi'(g) \quad \text{and} \quad 1_K,$$

respectively. Moreover it is immediate that the G-action on $\mathcal{F}$ "respects" the K-algebra operations:

$$(c\phi+c'\phi')\circ g = c(\phi\circ g) + c'(\phi'\circ g),\ (\phi\phi')\circ g = (\phi\circ g)(\phi'\circ g),\ \mathbf{1}\circ g = \mathbf{1}, \qquad (1.2a)$$

for all $c,c' \in K$, $\phi,\phi' \in \mathcal{F}$ and $g \in G$. If G acts on $\mathcal{M}$ on the right, then it acts on $\mathcal{F}$ on the left by the rule

$$(g\circ\phi)(m) = \phi(mg);$$

this action satisfies the equations analogous to (1.2a). Often these actions of G on $\mathcal{F}$ are changed to the other side by using the inverse e.g. here one could make G act on the right, defining $\phi\circ g$ by

$$(\phi\circ g)(m) = \phi(mg^{-1}).$$

(1.3) Now let m,n be positive integers, and K an infinite field. The groups $G_m = GL_m(K)$ and $G_n = GL_n(K)$ act, respectively on the left and right, on the set $\mathcal{M} = \mathcal{M}_{m,n}$ of all $m \times n$ matrices over K. So G_m, G_n act, respectively on the right and left, on the set $\mathcal{F}$ of all maps $\phi : \mathcal{M} \to K$. It is clear also that the action on $\mathcal{F}$ of any $g \in G_n$ commutes with that of any $h \in G_m$; we express this by saying that $\mathcal{F}$ is a (G_m,G_n)-*bimodule.*

Of special interest are the *coordinate maps* $c_{\mu\nu}$; for given $\mu \in \mathbf{m}$, $\nu \in \mathbf{n}$ one defines $c_{\mu\nu} \in \mathcal{F}$ by

$$c_{\mu\nu}(M) = M_{\mu\nu},$$

i.e. $c_{\mu\nu}$ takes each $M \in \mathcal{M}$ to its (μ,ν)-coefficient $M_{\mu\nu}$. Easy calculations show that $g \in G_n$ and $h \in G_m$ act on $c_{\mu\nu}$ to give

$$g \circ c_{\mu\nu} = \sum_{\nu' \in \mathbf{n}} g_{\nu'\nu}\, c_{\mu\nu'} \qquad (1.3a)$$

and

$$c_{\mu\nu} \circ h = \sum_{\mu' \in \mathbf{m}} h_{\mu\mu'}\, c_{\mu'\nu}, \qquad (1.3b)$$

respectively.

(1.4) Define

$$A(m \mid n) := K[c_{\mu\nu} : \mu \in \mathbf{m}, \nu \in \mathbf{n}],$$

the K-subalgebra of $\mathcal{F}$ generated by the mn maps $c_{\mu\nu}$. Since K is infinite, the $c_{\mu\nu}$ are algebraically independent over K; we may, if we like, regard the elements of $A(m \mid n)$ simply as polynomials over K in mn independent "variables" $c_{\mu\nu}$. From the standpoint of algebraic geometry, $A(m \mid n)$ is the coordinate ring of an affine space of dimension mn. From the standpoint of the representation theory of the general linear groups, the important thing is that $A(m \mid n)$ is closed to the actions of G_m and G_n; this follows from (1.3a), (1.3b), together with (1.2a) and its "left hand" analogue. There is a decomposition

$$A(m \mid n) = \sum_{r \geq 0}{}^{\oplus} A_r(m \mid n), \tag{1.4a}$$

where $A_0(m \mid n) = K.\mathbf{1}$, while, for each positive integer r, $A_r(m \mid n)$ is the K-space generated by all those elements of $A(m \mid n)$ which, considered as polynomials in the $c_{\mu\nu}$, are homogeneous of degree r. Clearly $A_r(m \mid n)$ is spanned, as K-space, by the set of all *monomials of degree* r

$$c_{i,j} = c_{i_1 j_1} c_{i_2 j_2} \cdots c_{i_r j_r}, \tag{1.4b}$$

where i runs over the set I(m,r) of all vectors or *indices* $(i_1, \ldots, i_r)$ with entries $i_\rho \in \mathbf{m}$, and j runs independently over the corresponding set I(n,r). (It is often useful to identify $i = (i_1, \ldots, i_r)$ with the map $i : \mathbf{r} \to \mathbf{m}$ which takes each $\rho \in \mathbf{r}$ to $i_\rho \in \mathbf{m}$. For this reason many authors use the notations $\mathbf{m}^{\mathbf{r}}$, $\mathbf{n}^{\mathbf{r}}$ instead of I(m,r), I(n,r), respectively.) Then from (1.3a) and (1.3b) follow that $g \in G_n$ and $h \in G_m$ act on $c_{i,j}$ by

$$g \circ c_{i,j} = \sum_{t \in I(n,r)} g_{t,j}\, c_{i,t} \tag{1.4c}$$

and

$$c_{i,j} \circ h = \sum_{s \in I(m,r)} h_{i,s}\, c_{s,j}, \tag{1.4d}$$

where we use the notations

$$g_{t,j} = g_{t_1 j_1} \cdots g_{t_r j_r} \text{ and } h_{i,s} = h_{i_1 s_1} \cdots h_{i_r s_r}. \tag{1.4e}$$

For example,

$$g \circ c_{i,j} = g \circ \left(c_{i_1 j_1} \cdots c_{i_r j_r}\right)$$

$$= \prod_{\rho \in \mathbf{r}} \left(g \circ c_{i_\rho j_\rho}\right)$$

$$= \prod_{\rho \in \mathbf{r}} \left(\sum_{t_\rho \in \mathbf{n}} g_{t_\rho j_\rho}\, c_{i_\rho t_\rho}\right) \text{ by (1.2a), (1.3a);}$$

and when this product is expanded, it gives the sum on the right side of (1.4c).

It follows from (1.4c) and (1.4d) that $A_r(m \mid n)$ is a (G_n, G_m)-sub-bimodule of $\mathfrak{F}$, for all $r > 0$. The same is true of $A_0(m \mid n)$, since $g \circ \mathbf{1} = \mathbf{1} = \mathbf{1} \circ h$ for all $g \in G_n$, $h \in G_m$.

From now on we keep m, n, r and K fixed. The Schur module $L_\lambda(K)$ (see (1.1)) will be constructed in (2.7) as a submodule of the left G_n-module $A_r(m \mid n)$.

(1.5) The symmetric group $P(r) = \mathrm{Sym}(\mathbf{r})$ pervades this work, not by any direct action on $A_r(m \mid n)$, but through its action on the index sets $I(m,r)$, $I(n,r)$ by "place permutations". This is a right action, as follows: if $\pi \in P(r)$ then

$$i\pi := (i_{\pi(1)}, \ldots, i_{\pi(r)}),\ j\pi := (j_{\pi(1)}, \ldots, j_{\pi(r)}),$$

for all $i \in I(m,r)$, $j \in I(n,r)$. Hence $P(r)$ acts on $I(m,r) \times I(n,r)$ by

$$(i,j)\pi = (i\pi, j\pi).$$

The symbol ~ shall mean "is in the same P(r)-orbit as", e.g. $i \sim i'$ means there exists $\pi \in P(r)$ such that $i\pi = i'$; and $(i,j) \sim (i',j')$ means there exists $\pi \in P(r)$ such that $i\pi = i'$ and $j\pi = j'$. An example of this notation is seen in the following easy (but important!) lemma:

Lemma (1.5a). *Let* $i,i' \in I(m,r)$ *and* $j,j' \in I(n,r)$. *Then the monomials* $c_{i,j}$ *and* $c_{i',j'}$ *are equal if and only if* $(i,j) \sim (i',j')$.

The number of *distinct* monomials of degree r in mn independent variables

$$c_{\mu\nu}\ (\mu \in \mathbf{m}, \nu \in \mathbf{n})$$

is $\binom{mn+r-1}{r}$, hence this number equals $\dim_K A_r(m \mid n)$. By (1.5a), it is also the number of P(r)-orbits in $I(m,r) \times I(n,r)$.

(1.6) The next move requires another basis for $A_r(m \mid n)$. The elements of this basis are certain polynomials in the $c_{\mu\nu}$ called *bideterminants*. To define these, we must introduce *shapes* and *tableaux*.

Definitions. A *shape* (or *diagram*) is just a finite, non-empty subset λ of $\mathbf{N} \times \mathbf{N}$. The number of elements of λ is denoted $|\lambda|$. A λ-*tableau* is a map

$$T : \lambda \to S,$$

S being an arbitrary set. If $T_{a,b}$ denotes the image under T of an element (a,b) of λ, we can depict T as a kind of "incomplete matrix" $(T_{a,b})_{(a,b)\in\lambda}$, leaving blank spaces for the places $(a,b) \in \mathbf{N} \times \mathbf{N}$ which do not belong to λ. For example if

$$\lambda = \{(1,1), (1,2), (1,4), (2,4), (3,2), (3,4)\},$$

then the λ-tableau T appears as shown.

T =

$T_{1,1}$	$T_{1,2}$		$T_{1,4}$
			$T_{2,4}$
	$T_{3,2}$		$T_{3,4}$

All the shapes we use will have r elements, and it is convenient to choose, for each such shape λ, a fixed bijection $T^\lambda : \lambda \to \mathbf{r}$, and call this the *basic λ-tableau*. Then each element ρ of $\mathbf{r}$ appears exactly once in the "matrix" representing T^λ. For any $a,b \in \mathbf{N}$ define the *a-th row* and the *b-th column* of T^λ by

$$\mathcal{R}_a = \mathcal{R}_a(T^\lambda) = \{T^\lambda_{a,n} \mid n \in \mathbf{N}\}$$

and

$$\mathcal{C}_b = \mathcal{C}_b(T^\lambda) = \{T^\lambda_{n,b} \mid n \in \mathbf{N}\},$$

respectively. These are subsets of $\mathbf{r}$, which may be empty; they determine two partitions of $\mathbf{r}$, namely

$$\mathbf{r} = \mathcal{R}_1 \cup \mathcal{R}_2 \cup \ldots = \mathcal{C}_1 \cup \mathcal{C}_2 \cup \ldots .$$

Example. Let $\lambda = \{(1,1), (1,2), (1,4), (2,4), (3,2), (3,4)\}$. Then $r = 6$, and we could take as basic λ-tableau

$T^\lambda =$

1	4		6
			3
	2		5

The rows are $\mathfrak{R}_1 = \{1,4,6\}$, $\mathfrak{R}_2 = \{3\}$, $\mathfrak{R}_3 = \{2,5\}$, the columns are $\mathfrak{C}_1 = \{1\}$, $\mathfrak{C}_2 = \{4,2\}$, $\mathfrak{C}_3 = \varnothing$, $\mathfrak{C}_4 = \{6,3,5\}$, and $\mathfrak{R}_a = \varnothing = \mathfrak{C}_b$ for all $a > 3$, $b > 4$.

Returning to the general case, the basic λ-tableau T^λ determines two subgroups of $P(r)$, namely

$$R(T^\lambda) = P(\mathfrak{R}_1)P(\mathfrak{R}_2)\ \dots \qquad (1.6a)$$

and

$$C(T^\lambda) = P(\mathfrak{C}_1)P(\mathfrak{C}_2)\ \dots, \qquad (1.6b)$$

called the *row-stabilizer* and *column-stabilizer* of T^λ, respectively. In these definitions, $P(S)$ denotes, for any subset S of $\mathbf{r}$, the subgroup of $P(r)$ consisting of all $\pi \in P(r)$ which leave fixed every element of $\mathbf{r}\backslash S$.

Note. The sets $\mathfrak{R}_a$, $\mathfrak{C}_b$ and the groups $R(T^\lambda)$, $C(T^\lambda)$ depend on the choice of basic tableau $T^\lambda : \lambda \to \mathbf{r}$. If $S^\lambda : \lambda \to \mathbf{r}$ is another bijective λ-tableau, then there is an element $\beta \in P(r)$ such that $S^\lambda = \beta T^\lambda$. One verifies easily that

$$\mathfrak{R}_a(T^\lambda) = \beta\mathfrak{R}_a(S^\lambda),$$

$$\mathfrak{C}_b(T^\lambda) = \beta\mathfrak{C}_b(S^\lambda),$$

$$R(T^\lambda) = \beta^{-1}R(S^\lambda)\beta,$$

and

$$C(T^\lambda) = \beta^{-1}C(S^\lambda)\beta.$$

In practice we often write $\mathfrak{R}_a(\lambda)$, $R(\lambda)$, etc. for $\mathfrak{R}_a(T^\lambda)$, $R(T^\lambda)$, etc., if the basic λ-tableau T^λ can be assumed known.

(1.7) Suppose $i \in I(m,r)$, $j \in I(n,r)$, and that $T^\lambda : \lambda \to \mathbf{r}$ has been chosen as basic λ-tableau for a given shape λ with $|\lambda| = r$. Define the λ-tableaux

$$T_i^\lambda : \lambda \to \mathbf{m}, \; T_j^\lambda : \lambda \to \mathbf{n}$$

to be the composites of T^λ with $i : \mathbf{r} \to \mathbf{m}$, $j : \mathbf{r} \to \mathbf{n}$, respectively. In the example of the last section, these λ-tableaux are

$$T_i^\lambda = \begin{array}{|c|c|c|c|} \hline i_1 & i_4 & & i_6 \\ \hline & & & i_3 \\ \hline & i_2 & & i_5 \\ \hline \end{array} \qquad T_j^\lambda = \begin{array}{|c|c|c|c|} \hline j_1 & j_4 & & j_6 \\ \hline & & & j_3 \\ \hline & j_2 & & j_5 \\ \hline \end{array}$$

respectively. Thus T_i^λ displays the entries i_ρ of i "according to T^λ"; the usual display $i = (i_1, i_2, ..., i_r)$ could be identified with T_i^α, where $T^\alpha = (1, 2, ..., r)$ is a basic tableau for the shape $\alpha = \{(1,1), (1,2), ..., (1,r)\}$.

Definition. With the notation above, let

$$(T_i^\lambda : T_j^\lambda) := \sum_{\pi \in C(\lambda)} (-1)^\pi \, c_{i\pi,j}. \tag{1.7a}$$

This is an element of $A_r(m \mid n)$, called a *bideterminant* of shape λ. In (1.7a) we have written $C(\lambda)$ for $C(T^\lambda)$, and $(-1)^\pi$ for the sign of the permutation π.

Example 1. Let $\beta = \{(1,1), (2,1), ..., (r,1)\}$. Take the basic β-tableau T^β as shown. Clearly $C(T^\beta) = P(r)$, hence

$$(T_i^\beta : T_j^\beta) = \sum_{\pi \in P(r)} (-1)^\pi c_{i\pi,j}$$

is the ordinary determinant of the $r \times r$ matrix $(c_{i_\rho j_\sigma})$ $\rho, \sigma \in \mathbf{r}$.

$$T^{\beta} = \begin{array}{|c|} \hline 1 \\ 2 \\ \cdot \\ \cdot \\ \cdot \\ r \\ \hline \end{array}\ , \quad (T_i^{\beta} : T_j^{\beta}) = \begin{bmatrix} i_1 & j_1 \\ i_2 & j_2 \\ \cdot & \cdot \\ \cdot & \cdot \\ \cdot & \cdot \\ i_r & j_r \end{bmatrix} = \det(c_{i_\rho j_\sigma}).$$

In general

$$(T_i^{\lambda} : T_j^{\lambda}) = \Gamma_1 \Gamma_2 \ldots, \tag{1.7b}$$

where for each column $\mathbf{C}_a$ of T^{λ}, Γ_a is the determinant of the matrix (c_{i_ρ, j_σ}) whose rows and columns are indexed by elements $\rho, \sigma \in \mathbf{C}_a(T^{\lambda})$. This follows quite easily from Definition (1.7a) and the fact that $C(\lambda) = C(T^{\lambda})$ is the direct product of the groups $P(\mathbf{C}_1)$, $P(\mathbf{C}_2)$, ...; see (1.6b). Notice that Γ_a is the bideterminant

$$\begin{bmatrix} i_u & j_u \\ i_v & j_v \\ \cdot & \cdot \\ \cdot & \cdot \\ \cdot & \cdot \\ i_w & j_w \end{bmatrix} = \begin{vmatrix} c_{i_u j_u} & c_{i_u j_v} & \cdots & c_{i_u j_w} \\ c_{i_v j_u} & c_{i_v j_v} & \cdots & c_{i_v j_w} \\ \cdots & \cdots & & \cdots \\ c_{i_w j_u} & c_{i_w j_v} & \cdots & c_{i_w j_w} \end{vmatrix} \tag{1.7c}$$

formed from the a-th columns of T_i^{λ}, T_j^{λ}. Here u, v, ..., w are the elements of the a-th column of the basic tableau T^{λ}, taken in order, reading from the top.

Example 2. Taking λ, i, j as in the example given at the beginning of this section, we find

$$(T_i^{\lambda} : T_j^{\lambda}) = c_{i_1 j_1} \begin{vmatrix} c_{i_4 j_4} & c_{i_4 j_2} \\ c_{i_2 j_4} & c_{i_2 j_2} \end{vmatrix} \begin{vmatrix} c_{i_6 j_6} & c_{i_6 j_3} & c_{i_6 j_5} \\ c_{i_3 j_6} & c_{i_3 j_3} & c_{i_3 j_5} \\ c_{i_5 j_6} & c_{i_5 j_3} & c_{i_5 j_5} \end{vmatrix}.$$

Example 3. Take $\alpha = \{(1,1),(1,2), ..., (1,r)\}$ and $T^{\alpha} = (1,2, ..., r)$. Since all the (non-empty) columns of T^{α} have length 1, we have for any

$i \in I(m,r)$, $j \in I(n,r)$

that

$$(T_i^\alpha : T_j^\alpha) = c_{i_1 j_1} c_{i_2 j_2} \dots c_{i_r j_r} = c_{i,j}.$$

Thus any monomial $c_{i,j}$ is a bideterminant of shape α.

Properties of bideterminants. The following properties will be used frequently:

(1.7d) $(T_i^\lambda : T_j^\lambda) = 0$ if and only if T_i^λ, or T_j^λ, has equal entries at distinct places ρ, σ in the same column of T^λ. For this is the necessary and sufficient condition that, for some column $\mathbf{C}_a$ of T^λ, the determinant (1.7c) should have two equal rows or two equal columns.

(1.7e) If $\pi \in C(T^\lambda)$ then

$$(T_{i\pi}^\lambda : T_j^\lambda) = (-1)^\pi (T_i^\lambda : T_j^\lambda) = (T_i^\lambda : T_{j\pi}^\lambda).$$

This comes at once from the Definition (1.7a).

(1.7f) From the Definition (1.7a) it is clear that

$$(S_i^\mu : S_j^\mu) = (T_i^\lambda : T_j^\lambda),$$

where μ is any shape having a basic tableau S^μ whose columns $\mathbf{C}_1(S^\mu)$, $\mathbf{C}_2(S^\mu)$ ... are the same subsets of $\mathbf{r}$ as the columns $\mathbf{C}_1(T^\lambda)$, $\mathbf{C}_2(T^\lambda)$, ... of T^λ, although possibly in a different order. For in this case the column-stabilizers $C(S^\mu)$ and $C(T^\lambda)$ coincide (see (1.6b)). As an example, consider the basic tableaux

$T^\lambda =$

1	4		6
			3
	2		5

$S^\mu =$

3	2	1
5	4	
6		

Then $(S_i^\mu : S_j^\mu) = (T_i^\lambda : T_j^\lambda)$ for any $i \in I(m,6)$, $j \in I(n,6)$.

(1.8) A *partition* λ *of* r is by definition a sequence $(\lambda_1, \lambda_2, ...)$ of non-negative integers $\lambda_1, \lambda_2, ...$ which satisfies

$$\lambda_1 \geq \lambda_2 \geq ... \text{ and } \sum_{\nu \in N} \lambda_\nu = r.$$

We write $\lambda \vdash r$ to indicate that λ is a partition of r. It is usual to omit the zero terms in $(\lambda_1, \lambda_2, ...)$, so that, for example, the partition (4, 1, 1, 0, 0, ...) of 6 is written (4, 1, 1) [alternative notations are (411), or (41^2)]. The non-zero terms λ_a of the sequence $(\lambda_1, \lambda_2, ...)$ are called the *parts* of λ.

Associated to a partition λ is its shape

$$\lambda := \{(\mu,\nu) \in N \times N \mid 1 \leq \nu \leq \lambda_\mu\}.$$

We use the same symbol λ to denote both the partition and its shape. The conjugate $\tilde{\lambda}$ of λ is that partition of r whose shape is the "transpose" of the shape of λ. $\tilde{\lambda} = (\tilde{\lambda}_1, \tilde{\lambda}_2, ...)$ can also be described as follows: for each a = 1,2, ...,

$$\tilde{\lambda}_a = (\text{number of rows of } \lambda \text{ of lengths} \geq a)$$

$$= \text{length of column a of the shape } \lambda.$$

In particular

$$\tilde{\lambda}_1 = \text{number of parts of } \lambda$$

$$= \text{length of column 1 of the shape } \lambda.$$

Example 1. Let α be the partition (r) = (r, 0, 0, ...). Then the

$$\textit{shape } \alpha = \{(1,1), (1,2), ..., (1,r)\}$$

(see (1.7), Example 3). The partition conjugate to α is

$$\tilde{\alpha} = \beta = (1, 1, ..., 1) = (1^r),$$

and its shape is

$$\beta = \{(1,1), (2,1), ..., (r,1)\}$$

(see (1.7), Example 1).

Example 2. If T^λ is a basic tableau for (the shape λ of) a partition λ, then the "transpose" $S^{\tilde{\lambda}}$ of T^λ is a basic tableau for $\tilde{\lambda}$; for example

$$T^\lambda = \begin{matrix} 1 & 2 & 3 & 4 \\ 5 & 6 & 7 & \\ 8 & 9 & & \end{matrix} \quad , \qquad S^{\tilde{\lambda}} = \begin{matrix} 1 & 5 & 8 \\ 2 & 6 & 9 \\ 3 & 7 & \\ 4 & & \end{matrix} \quad ,$$

are basic tableaux for $\lambda = (4\ 3\ 2)$, $\tilde{\lambda} = (3\ 3\ 2\ 1) = (3^2\ 2\ 1)$, respectively.

Notation. From now on, tableaux will be written as above, i.e. without dividing lines.

(1.9) Let $\lambda \vdash r$, $i \in I(m,r)$, $j \in I(n,r)$. From the definition (1.7a) of bideterminant, and the formulae (1.4c) and (1.4d), follow at once

$$g \circ (T_i^\lambda : T_j^\lambda) = \sum_{t \in I(n,r)} g_{t,j}(T_i^\lambda : T_t^\lambda) \tag{1.9a}$$

and

$$(T_i^\lambda : T_j^\lambda) \circ h = \sum_{s \in I(m,r)} h_{i,s}(T_s^\lambda : T_j^\lambda), \tag{1.9b}$$

for all $g \in G_n$, $h \in G_m$. These formulae show that the K-space B_λ spanned by all $(T_i^\lambda : T_j^\lambda)$ of given shape λ, as i, j range over I(m,r), I(n,r), respectively, is a (G_n, G_m)-submodule (or "sub-bimodule") of $A_r(m \mid n)$. However the relations between the B_λ, for different $\lambda \vdash r$, are difficult to describe, and it is better to work with the spaces

$$A_\lambda := \sum_{\mu \le \lambda} B_\mu = \sum_{\mu \le \lambda} \sum_{i,j} K \cdot (T_i^\mu : T_j^\mu), \tag{1.9c}$$

where $\le$ denotes the following total order on the set of all partitions of r: by definition, $\mu \le \lambda$ if $\mu = \lambda$, or if $\tilde{\mu} > \tilde{\lambda}$ in the lexicographic order of partitions, i.e. if there exists $a \in \mathbf{r}$ such that

$$\tilde{\mu}_1 = \tilde{\lambda}_1, \ \ldots, \ \tilde{\mu}_{a-1} = \tilde{\lambda}_{a-1}, \ \tilde{\mu}_a > \tilde{\lambda}_a.$$

In this order, (r) is the "largest" partition of r, and (1^r) is the "smallest"; if r = 5, we have

$$(5) > (41) > (32) > (31^2) > (2^21) > (21^3) > (1^5).$$

It is clear that each A_λ is a (G_n, G_m)-submodule of $A_r(m \mid n)$, and that $\lambda \geq \mu$ implies $A_\lambda \geq A_\mu$. Moreover $A_{(r)} = A_r(m \mid n)$, because every monomial $c_{i,j}$ can be written as a bideterminant of shape (r) (see (1.7), Example 3). So if

$$\lambda(1), \lambda(2), ..., \lambda(t)$$

are the partitions of r written in decreasing order, so that $\lambda(1) = (r)$, $\lambda(t) = (1^r)$, there is a chain of (G_n, G_m)-submodules of $A_r(m \mid n)$,

$$A_r(m \mid n) = A_{\lambda(1)} \geq A_{\lambda(2)} \geq ... \geq A_{\lambda(t)} \geq \{0\}, \tag{1.9d}$$

which we may call the canonical filtration of $A_r(m \mid n)$.

Definition 1. Let $\lambda \vdash r$, $i \in I(m,r)$. Then the λ-tableau T_i^λ is *standard* if the entries in each row increase (≤) from left to right, and the entries in each column increase strictly (<) from top to bottom. (Many authors would interchange "row" and "column" in this definition.)

Definition 2. Let $\lambda \vdash r$, $i \in I(m,r)$, $j \in I(n,r)$. Then the bideterminant $(T_i^\lambda : T_j^\lambda)$ is *standard*, if both T_i^λ and T_j^λ are standard.

Standard tableaux were introduced by A Young in his work on substitutional analysis ([Y, p.258]). He found that, for a given partition λ of r, the standard bijective λ-tableaux $S : \lambda \to \mathbf{r}$ can serve to parametrize a basis for an irreducible (complex) P(r)-module. The next theorem shows how standard bideterminants provide a basis for $A_r(m \mid n)$, and indeed for each of the spaces $A_{\lambda(\tau)}$ in the filtration (1.9d).

Theorem 1 (Mead (1972); Doubilet, Rota, Stein (1974)). *Let* m, n, r *and* K *be given, and let* λ *be a partition of* r. *Then the set*

$$\left\{(T_h^\mu : T_k^\mu) \mid \mu \leq \lambda,\ T_h^\mu, T_k^\mu \textit{ both standard } (h \in I(m,r),\ k \in I(n,r))\right\} \tag{1.9e}$$

is a K*-basis of the space* A_λ *defined in (1.9c).*

Taking $\lambda = (r)$, we get the basis of $A_r(m \mid n)$ promised in (1.6).

(1.9f) *The set of all standard bideterminants*

$$(T_h^\lambda : T_k^\lambda) \text{ for } \lambda \vdash r,\ h \in I(m,r),\ k \in I(n,r)$$

is a K-*basis of* $A_r(m \mid n)$.

Another corollary of Theorem 1 is:

(1.9g) *Let* $\tau \in \mathbf{t} = \{1,\dots,t\}$. *Then* $A_{\lambda(\tau)}/A_{\lambda(\tau+1)}$ *has as* K-*basis the set of all*

$$(T_h^{\lambda(\tau)} : T_k^{\lambda(\tau)}) + A_{\lambda(\tau+1)}$$

with $(T_h^{\lambda(\tau)} : T_k^{\lambda(\tau)})$ *standard and* $h \in I(m,r)$, $k \in I(n,r)$.

Remark. If $\tau = t$, we take $A_{\lambda(t+1)} = \{0\}$ in (1.9g).

We end this section with yet another corollary of Theorem 1.

(1.9h) *Let* $\lambda \vdash r$. *Then the subspace* A_λ *of* $A_r(m \mid n)$ *(see (1.9c)) is non-zero if and only if* $\tilde{\lambda}_1 \leq \min(m,n)$.

Proof. The entries in any column of a standard tableau must be distinct. Hence if T_h^λ, $h \in I(m,r)$, is standard, then $\tilde{\lambda}_1$ (which is the length of the longest column of T_h^λ) cannot exceed m. Similarly if there exists a standard tableau T_k^λ, $k \in I(n,r)$, then $\tilde{\lambda}_1 \leq n$. But Theorem 1 shows that if $A_\lambda \neq 0$, there exist standard tableaux T_h^μ, T_k^μ for some $\mu \leq \lambda$. This proves that $\tilde{\mu}_1 \leq \min(m,n)$; but $\tilde{\mu} \leq \tilde{\lambda}$ implies $\tilde{\mu}_1 \geq \tilde{\lambda}_1$, hence $\tilde{\lambda}_1 \leq \min(m,n)$.

Conversely if $d = \tilde{\lambda}_1 \leq \min(m,n)$, there is an index

$$\ell \in I(d,r) \leq I(m,r) \cap I(n,r)$$

such that

$$T_\ell^\lambda = \begin{matrix} 1 & 1 & \dots & 1 \\ 2 & 2 & \dots & 2 \\ d & \dots & d & \end{matrix} ;$$

this is the *canonical* λ*-tableau*, which will play an important part in Chapter II. Since T_ℓ^λ is clearly standard, the set (1.9e) contains the standard bideterminant $(T_\ell^\lambda : T_\ell^\lambda)$, and this proves that $A_\lambda \neq 0$.

(1.10) This section and the next are devoted to the proof of one half of Theorem I, viz.:

(1.10a) *Let* $\lambda \vdash r$ *be given. Then the set (1.9e)* K*-spans the subspace* A_λ *of* $A_r(m \mid n)$.

The proof of the rest of Theorem I, viz. that (1.9e) is a K-linearly independent set, will be given in Chapter II. In their essentials both proofs are the same as those given by Mead ([M, Theorems 1 and 2]).

The proof of (1.10a) goes in several stages, and gives in fact this slightly stronger version:

(1.10b) *Let* $\lambda \vdash r$, $i \in I(m,r)$, $j \in I(n,r)$. *Then* $(T_i^\lambda : T_j^\lambda)$ *can be written as a* **Z***-linear combination of the elements of the set (1.9e).*

Here $A_r(m \mid n)$ is regarded as **Z**-module, by defining za to be $(z \cdot 1_K)a$, for any $z \in \mathbf{Z}$, $a \in A_r(m \mid n)$. ($A_r(m \mid n)$ is not a *free* **Z**-module, unless the characteristic of K is zero.) Clearly (1.10b) implies (1.10a), since a **Z**-linear combination of given elements of $A_r(m \mid n)$, is also a K-linear combination of these elements.

Each basic λ-tableau T^λ determines an order on the set **r**, by going down each column of T^λ in succession, starting from the left. We call this the T^λ-order on **r**. For example if $\lambda = (421)$ and

$$T^\lambda = \begin{matrix} 1 & 2 & 3 & 4 \\ 5 & 6 & & \\ 7 & & & \end{matrix} \quad ,$$

then the T^λ-order of **7** is 1, 5, 7, 2, 6, 3, 4. Next we order the set of all λ-tableaux T_i^λ, $i \in I(m,r)$, by the rule: let $p,i \in I(m,r)$, then $T_p^\lambda < T_i^\lambda$ if at the "first"

$\rho \in \mathbf{r}$ where $p_\rho \neq i_\rho$, there holds $p_\rho < i_\rho$. Here "first" refers to the T^λ-order just described. The first stage in the proof of (1.10b) is the following lemma.

Lemma (1.10c). *Let* $i \in I(m,r)$. *If* T_i^λ *is not standard, then there exist integers* $z_{i,p}$, *and a partition* $\zeta < \lambda$, *such that for any* $j \in I(n,r)$ *there holds an equation*

$$(T_i^\lambda : T_j^\lambda) = \sum_p z_{i,p} (T_p^\lambda : T_j^\lambda) + L_j, \tag{1.10d}$$

where the sum is over all $p \in I(m,r)$ *such that* $T_p^\lambda < T_i^\lambda$, *and* L_j *is a* **Z**-*linear combination of bideterminants of shape* ζ.

Remark. The proof of (1.10c) will determine the integers $z_{i,p}$ and the partition ζ in a way which is independent of the field K.

Proof of (1.10c). We may assume that the entries in each column of T_i^λ are distinct, for otherwise $(T_i^\lambda : T_j^\lambda) = 0$ for all T_j^λ (see (1.7d)). And we may even assume that the entries in each column increase strictly from top to bottom, for there must exist some $\pi \in C(T^\lambda)$ such that $T_{i\pi}^\lambda$ does have this property, and we may replace T_i^λ by $T_{i\pi}^\lambda$ since by (1.7e)

$$(T_{i\pi}^\lambda : T_j^\lambda) = (-1)^\pi (T_i^\lambda : T_j^\lambda)$$

for all T_j^λ.

Since T_i^λ is not standard, there is some $\rho \in \mathbf{r}$ having σ immediately to its right in T^λ, such that $i_\rho > i_\sigma$. Suppose

$$\rho \in C_a(T^\lambda) = C_a, \quad \sigma \in C_{a+1}(T^\lambda) = C_{a+1}.$$

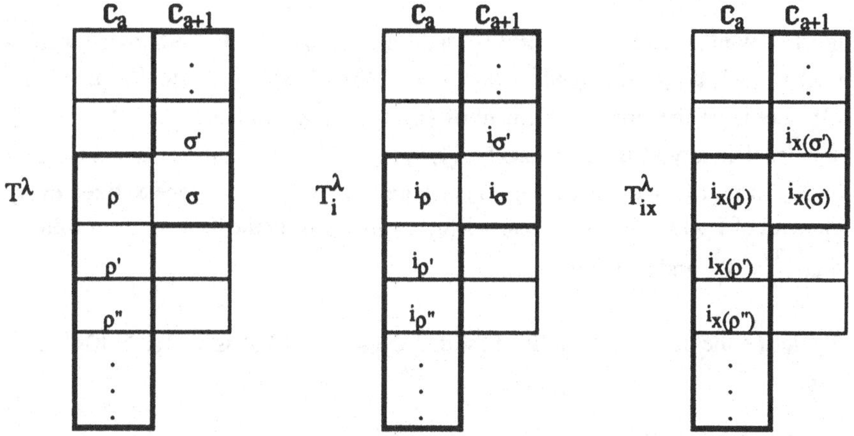

Let $S = \{\ldots, \rho'', \rho', \rho, \sigma, \sigma', \sigma'', \ldots\}$, where $\rho', \rho'', \ldots$ are the elements of $\mathbf{C}_a$ which are below ρ (if any such exist), and $\sigma', \sigma'', \ldots$ are the elements of $\mathbf{C}_{a+1}$ which are above σ (if any such exist).

Since $i_\rho > i_\sigma$ and the columns of T_i^λ are strictly increasing, there holds

$$\ldots > i_{\rho''} > i_{\rho'} > i_\rho > i_\sigma > i_{\sigma'} > \ldots . \tag{1}$$

Recall that $P(S)$ denotes the group of all $\pi \in P(r)$ which fix all elements of $\mathbf{r}\backslash S$, and that $C(T^\lambda) = P(\mathbf{C}_1)\, P(\mathbf{C}_2) \ldots$, where

$$\mathbf{C}_1, \mathbf{C}_2, \ldots \tag{2}$$

are the columns of T^λ. If $x' \in P(S)$, it is easy to see that the coset

$$x'\,(P(S) \cap C(T^\lambda))$$

contains exactly one element x which satisfies

$$\ldots > i_{x(\rho'')} > i_{x(\rho')} > i_{x(\rho)} \quad \text{and} \quad i_{x(\sigma)} > i_{x(\sigma')} > \ldots . \tag{3}$$

Hence the set X of all $x \in P(S)$ satisfying (3), is a transversal of the left cosets of $P(S) \cap C(T^\lambda)$ in $P(S)$. Moreover

$$\text{if } 1 \neq x \in X, \text{ then } T_{ix}^\lambda < T_i^\lambda. \tag{4}$$

To prove this, observe that the set

$$M = \{..., i_{x(\rho'')}, i_{x(\rho')}, i_{x(\rho)}\}$$

must contain at least one element from the set $\{i_\sigma, i_{\sigma'}, ...\}$. For otherwise, x must map $\{..., \rho'', \rho', \rho\}$ into itself, hence $x \in C(\lambda) \cap P(S)$, which implies $x = 1$, by (3). Let i_τ be the smallest element of $\{i_\sigma, i_{\sigma'}, ...\}$ which lies in M. By (1), i_τ is smaller than any of the elements $i_\rho, i_{\rho'}, i_{\rho''}, ...$. Hence i_τ is the smallest element of M, so that $i_\tau = i_{x(\rho)}$, by (3). Therefore $i_{x(\rho)} < i_\rho$. But since x fixes every element of r which is not in S, $i_\nu = i_{x(\nu)}$ for all $\nu < \rho$ (in the T^λ-order). It follows that $T^\lambda_{ix} < T^\lambda_i$, and so (4) is proved.

To define the partition ζ, notice first that the sets in (2) satisfy $|C_1| \geq |C_2| \geq ...$ and

$$\tilde{\lambda} = (|C_1|, |C_2|, ...). \tag{5}$$

Now replace the two sets C_a, C_{a+1} in (2), by the three sets $C_a \backslash S$, $C_{a+1} \backslash S$, S (some of these may be empty), leaving the other sets C_f ($f \neq a, a+1$) unchanged. Let

$$C'_1, C'_2, ... \tag{6}$$

be the resulting collection of sets, rearranged so that $|C'_1| \geq |C'_2| \geq ...$. Define ζ to be the partition of r such that

$$\tilde{\zeta} = (|C'_1|, |C'_2|, ...). \tag{7}$$

Since $|S| = |C_a| + 1$, while $|C_a \backslash S|$, $|C_{a+1} \backslash S|$ are both less than $|C_a|$, we can arrange (6) so that

$$C'_1 = C_1, ..., C'_b = C_b, C'_{b+1} = S,$$

where $C_1, ..., C_b$ are all the elements of (2) of cardinal $> |C_a|$. But then

$$|C'_{b+1}| > |C_{b+1}|,$$

and (5), (7) show that $\tilde{\zeta}$ exceeds $\tilde{\lambda}$ lexicographically. Therefore $\zeta < \lambda$, as required by Lemma (1.10c).

There is a basic ζ-tableau T^{ζ} whose columns are the set (6). Then

$$C(T^{\zeta}) = \prod_{f} P(C'_f)$$

$$= \left\{ \prod_{f \neq a, a+1} P(C_f) \right\} P(C_a \backslash S)\ P(C_{a+1} \backslash S)\ P(S).$$

Since $P(C_a \backslash S) \leq P(C_a)$, $P(C_{a+1} \backslash S) \leq P(C_{a+1})$, it follows that

$$C(T^{\zeta})\ C(T^{\lambda}) = P(S)C(T^{\lambda}).$$

Therefore the set X above is a transversal of the left cosets of $C(T^{\zeta}) \cap C(T^{\lambda})$ in $C(T^{\zeta})$. Let Y be any transversal of the right cosets of $C(T^{\zeta}) \cap C(T^{\lambda})$ in $C(T^{\lambda})$. Then Clausen's "Laplace Duality Theorem" (see Section 1.11) gives, for any $j \in I(n,r)$,

$$\sum_{x \in X} (-1)^x (T^{\lambda}_{ix} : T^{\lambda}_{j}) = \sum_{y \in Y} (-1)^y (T^{\zeta}_{i} : T^{\zeta}_{jy^{-1}}). \tag{8}$$

Write L_j for the right member of equation (8), and transfer to the right side all the terms of the left member, except the term with $x = 1$. The resulting equation

$$(T^{\lambda}_{i} : T^{\lambda}_{j}) = - \sum_{\substack{x \in X \\ x \neq 1}} (-1)^x (T^{\lambda}_{ix} : T^{\lambda}_{j}) + L_j \tag{9}$$

can be taken as the equation (1.10d) required by Lemma (1.10c), since by (4), $T^{\lambda}_{ix} < T^{\lambda}_{j}$ for all $x \in X$, $x \neq 1$. This completes the proof of Lemma (1.10c).

The rest of the proof of (1.10b) is relatively easy! First we demonstrate the following corollary of (1.10c).

Lemma (1.10e). *Let* $\lambda \vdash r$, $i \in I(m,r)$. *Then, for each* $h \in I(m,r)$ *with* T^{λ}_{h} *standard, there is an integer* $w_{i,h}$, *so that for any* $j \in I(n,r)$ *there holds*

$$(T^{\lambda}_{i} : T^{\lambda}_{j}) = \sum_{T^{\lambda}_{h}\,\text{std.}} w_{i,h} (T^{\lambda}_{h} : T^{\lambda}_{j}) + F_j, \tag{1.10f}$$

where F_j *is a* $\mathbb{Z}$*-linear combination of bideterminants of shapes* $< \lambda$.

Remark. It will be important, later on, that *the integers* $w_{i,h}$ *are independent of* j. There are also independent of K.

Proof of (1.10e). If (1.10e) is false, let T_i^λ be the smallest λ-tableau (in the ordering above) for which it fails. T_i^λ cannot be standard, for in that case (1.10e) would hold, taking $w_{i,h} = 1$ or 0, according as $i = h$ or $i \neq h$, and taking $F_j = 0$. So we may apply (1.10c). But since (1.10e) holds for each $T_p^\lambda < T_i^\lambda$, one deduces from (1.10d) that (1.10e) holds also for T_i^λ. This contradiction proves Lemma (1.10e).

It is clear that Lemma (1.10e) has a "right-hand" analogue, namely:

Lemma (1.10e'). *Let* $\lambda \vdash r$, $j \in I(n,r)$. *Then, for each* $k \in I(n,r)$ *with* T_k^λ *standard, there is an integer* $w'_{k,j}$, *so that for any* $i \in I(m,r)$ *there holds*

$$(T_i^\lambda : T_j^\lambda) = \sum_{T_k^\lambda \text{ std.}} w'_{k,j} (T_i^\lambda : T_k^\lambda) + F'_i , \qquad (1.10f')$$

where F'_i *is a* **Z**-*linear combination of bideterminants of shapes* $< \lambda$.

Lemma (1.10e') can be deduced from (1.10e), by using the K-algebra isomorphism

$$A(m \mid n) \to A(n \mid m)$$

which takes $c_{\mu\nu} \to c_{\nu\mu}$, for all $\mu \in \mathbf{m}$, $\nu \in \mathbf{n}$. We leave the details to the reader.

Proof of (1.10b). Combine (1.10f) with (1.10f'), and remember that for fixed j, (1.10f') holds for all $i \in I(m,r)$. We find

$$(T_i^\lambda : T_j^\lambda) = \sum_{h,k} w_{i,h}\, w'_{k,j} (T_h^\lambda : T_k^\lambda) + P \qquad (1.10g)$$

where the sum is over all standard T_h^λ, T_k^λ, and P is a **Z**-linear combination of bideterminants of shapes $< \lambda$. The proof of (1.10b) is now completed by induction on λ.

(1.11) Appendix to Chapter I

The theorem that the standard bideterminants span $A_r(m \mid n)$ has been called the *Straightening Formula* by G-C Rota and his collaborators ([DRS, p.198], [DKR, p.68]). The proofs in the works just cited, as well as the earlier proof of Mead (1972), use induction with respect to a linear ordering of λ-tableaux; the induction step is then based on an identity which expresses a non-standard bideterminant Δ as an integral combination of "smaller" ones of the same shape λ as Δ, together with bideterminants of shapes "smaller" than λ. In [M], [DRS] and [DKR] this identity is provided by a Laplace expansion of a suitable determinant. Recently M Clausen has given a very simple general formula which has as special cases the ordinary Laplace expansions of a determinant, and also the identity needed in the present work (see 1.10(8)). For the reader's convenience we state and prove Clausen's formula in the notation of these lectures.

Laplace Duality Theorem (1.11a) ([Cl', Theorem 3.1]; [Cl", p.46]). *Let λ,ζ be any shapes of order r, and let T^λ, T^ζ be basic tableaux for these shapes. Take any transversal X of the left cosets of*

$$C(T^\lambda) \cap C(T^\zeta)$$

in $C(T^\zeta)$, and any transversal Y of the right cosets of $C(T^\lambda) \cap C(T^\zeta)$ in $C(T^\lambda)$. Then for any $i \in I(m,r)$, $j \in I(n,r)$ there holds

$$\sum_{x\in X} (-1)^x (T^\lambda_{ix} : T^\lambda_j) = \sum_{y\in Y} (-1)^y (T^\zeta_i : T^\zeta_{jy^{-1}}). \tag{1.11b}$$

Proof of (1.11a). The set $Z = C(T^\zeta)C(T^\lambda)$ can be expressed in two ways as disjoint union of cosets, namely

$$Z = \bigcup_{x\in X} xC(T^\lambda)$$

$$= \bigcup_{y\in Y} C(T^\zeta)y.$$

Hence the left side of (1.11b) equals

$$\sum_{x\in X}\sum_{\pi\in C(\lambda)}(-1)^x(-1)^\pi c_{ix\pi,j}$$

$$=\sum_{z\in Z}(-1)^z c_{iz,j}$$

$$=\sum_{\sigma\in C(\zeta)}\sum_{y\in Y}(-1)^y(-1)^\sigma c_{i\sigma y,j}.$$

But since $c_{i\sigma y,j} = c_{i\sigma,jy^{-1}}$ (see (1.5a)), the last expression equals the right side of (1.11b).

II. Action of U^-_m on $A_r(m \mid n)$

(2.1) In this chapter we complete the proof of Theorem 1 (see Section (1.9)) by proving that

$$\{(T^\lambda_h : T^\lambda_k) \mid \lambda \vdash r,\ h \in I(m,r),\ k \in I(n,r),\ T^\lambda_h,\ T^\lambda_k \text{ both standard}\} \qquad (2.1a)$$

is a linearly independent subset of the K-space $A_r(m \mid n)$. Therefore, for given λ, the set (1.9e) is a linearly independent subset of A_λ.

It is convenient to dispose of the case m = 1 now. In this case, standard bideterminants $(T^\lambda_i : T^\lambda_j)$ occur only if $\lambda = (r)$, and they have the form

$$(1\ 1\ \ldots\ 1 : j_1\ j_2\ \ldots\ j_r) = c_{ij_1}\, c_{ij_2} \cdots c_{ij_r},$$

where $j_1 \le j_2 \le \ldots \le j_r$. Therefore (2.1a) consists of all the monomials of degree r in the variables $c_{11}, c_{12}, \ldots, c_{1n}$, with no monomial repeated. It is clear that this set is linearly independent. From now on in this chapter we assume $m \ge 2$.

We shall follow Mead (1972) and de Concini, Eisenbud, Procesi (1980), who base their proofs of (2.1a) on the (right) action on $A_r(m \mid n)$ of the group

$$U^- = U^-_m(K)$$

of all lower unitriangular matrices in $G_m = GL_m(K)$. This method gives, as a very useful byproduct, a K-basis of the space $A_r(m \mid n)^{U^-}$ of all U^-_m-invariant elements of $A_r(m \mid n)$. This leads to fundamental properties of the Schur modules L_λ, which are discussed in Sections (2.7)-(2.8).

(2.2) For each pair of distinct integers $p,q \in \mathbf{m}$, and each element $t \in K$, define the $m \times m$ matrix

$$u_{p,q}(t) = \begin{bmatrix} 1 & & & & & \\ & 1 & & & & \\ & & \cdot & & & \\ & & & \cdot & & \\ & & & & \cdot & \\ & t & & & 1 & \\ & & & & & 1 \end{bmatrix}, \tag{2.2a}$$

which has t in place (p,q), and is elsewhere identical with the identity matrix. The set

$$U_{p,q} = \{u_{p,q}(t) \mid t \in K\} \tag{2.2b}$$

is a subgroup of $G_m = GL_m(K)$, often called a *root subgroup*. It is not hard to see that $U^-_m(K)$ is generated by the set of all $U_{p,q}$, $1 \leq q < p \leq m$. For the rest of this section $\lambda \vdash r$ is fixed, and p,q are fixed elements of $\mathbf{m}$ such that $p \neq q$.

Let $i \in I(m,r)$, $j \in I(n,r)$ and $t \in K$. Put $h = u_{p,q}(t)$ in formula (1.9b):

$$(T_i^\lambda : T_j^\lambda) \circ h = \sum_{s \in I(m,r)} h_{i,s}\, (T_s^\lambda : T_j^\lambda). \tag{1}$$

The coefficient $h_{\mu,\nu}$ ($\mu,\nu \in \mathbf{m}$) of h is zero unless (μ,ν) lies in the set

$$B = \{(1,1), (2,2), ..., (m,m), (p,q)\}.$$

Hence the coefficient $h_{i,s} = h_{i_1 s_1} \dots h_{i_r s_r}$ in (1) is zero unless

$$(i_\rho, s_\rho) \in B, \text{ for all } \rho \in \mathbf{r}. \tag{2}$$

Assume now that (2) holds. Then $i_\rho = s_\rho$, for all $\rho \in \mathbf{r}$ which do not lie in the set

$$R = R(i,s) = \{\rho \in \mathbf{r} \mid (i_\rho, s_\rho) = (p,q)\}.$$

Clearly R is a subset of $D_p(i) = \{\rho \in \mathbf{r} \mid i_\rho = p\}$; conversely, with $i \in I(m,r)$ given, each subset R of $D_p(i)$ determines a unique element $s(R) \in I(m,r)$ such that (2) holds and $R = R(i,s)$. Namely $s = s(R)$ is given by

$$s_\rho = i_\rho \ (\rho \in \mathbf{r} \setminus R),\ s_\rho = q (\rho \in R), \tag{3}$$

i.e. s is obtained from i by replacing $i_\rho = p$ with q, at all places $\rho \in R$. From (3) we see that $h_{i,s} = t^{|R|}$ if $s = s(R)$. So (1) may be rewritten:

$$(T_i^\lambda : T_j^\lambda) \circ u_{p,q}(t) = \sum_R t^{|R|} (T_{s(R)}^\lambda : T_j^\lambda), \tag{4}$$

summed over all subsets R of $D_p(i)$.

Lemma (2.2c). *Formula (4) remains true, if the sum is restricted to the subsets of*

$$D_{p,q}^\lambda(i) = \{\rho \in \mathbf{r} \mid i_\rho = p, \textit{ and } i_\sigma \neq q \textit{ for all } \sigma \textit{ in the same column of } T^\lambda \textit{ as } \rho\}.$$

Proof. It will be enough to prove that $(T_{s(R)}^\lambda : T_j^\lambda)$ is zero, for any subset R of $D_p(i)$, which is not contained in $D_{p,q}^\lambda(i)$. Such a set R contains some $\rho \notin D_{p,q}^\lambda(i)$; so that there is some σ in the same column of T^λ as ρ, for which $i_\sigma = q$. But then $s = s(R)$ has $s_\rho = s_\sigma = q$, hence $(T_{s(R)}^\lambda : T_j^\lambda) = 0$ by (1.7d).

Example 1. Let $r = 7$, $m \geq 4$, $p = 3$, $q = 2$, and $\lambda = (4, 2, 1)$. Take the basic λ-tableau T^λ, below, and define $i \in I(m,r)$ by the λ-tableau T_i^λ shown.

$$T^\lambda = \begin{matrix} 1 & 2 & 3 & 4 \\ 5 & 6 & & \\ 7 & & & \end{matrix} \qquad T_i^\lambda = \begin{matrix} 1 & 2 & 3 & 4 \\ 3 & 3 & & \\ 4 & & & \end{matrix} .$$

Then $D_3(i) = \{3, 5, 6\}$ and $D_{3,2}^\lambda(i) = \{3, 5\}$.

Take some fixed $j \in I(n,r)$, and write $[T_s]$ for $(T_s^\lambda : T_j^\lambda)$. Now (4) and (2.2c) give, for all $t \in K$,

$$[T_i] \circ u_{3,2}(t) = [T_i] + t\begin{bmatrix} 1 & 2 & 2 & 4 \\ 3 & 3 & & \\ 4 & & & \end{bmatrix} + t\begin{bmatrix} 1 & 2 & 3 & 4 \\ 2 & 3 & & \\ 4 & & & \end{bmatrix} + t^2\begin{bmatrix} 1 & 2 & 2 & 4 \\ 2 & 3 & & \\ 4 & & & \end{bmatrix}.$$

Going back to the general case, (4) and (2.2c) give us the following formula: for all $\lambda \vdash r$, $i \in I(m,r)$, $j \in I(n,r)$, $p,q \in \mathbf{m}$ $(p \neq q)$ there holds

$$(T_i^\lambda : T_j^\lambda) \circ u_{p,q}(t) = X_0 + tX_1 + \ldots + t^d X_d, \text{ for all } t \in K. \tag{2.2d}$$

Here $d = |D_{p,q}^\lambda(i)|$, and

$$X_f = X_f(i,j) = \sum_{|R|=f} (T_{s(R)}^\lambda : T_j^\lambda) \tag{2.2e}$$

for all $f = 0,1,\ldots, m$; the sum in (2.2e) is over all subsets R of $D_{p,q}^\lambda(i)$ of cardinal f.

Example 2. Since $s(\varnothing) = i$ (see (3)), $X_0 = (T_i^\lambda : T_j^\lambda)$. Hence if $D_{p,q}^\lambda(i) = \varnothing$, then (2.2d) shows that $(T_i^\lambda : T_j^\lambda)$ is invariant to right action by all $u \in U_{p,q}$. Now assume $\tilde{\lambda}_1 \leq m$, and take $i = \ell(\lambda)$ to be the element of $I(m,r)$ such that $T_{\ell(\lambda)}^\lambda$ is the *canonical λ-tableau* mentioned in Section (1.9): it is defined by

$$\rho \in \mathcal{R}_a(T^\lambda) \Rightarrow \ell_\rho = a \quad (a = 1,2,\ldots). \tag{2.2f}$$

It is easy to see that $D_{p,q}^\lambda(\ell(\lambda)) = \varnothing$ for all $p,q \in \mathbf{m}$ such that $p > q$. Hence

$$(T_{\ell(\lambda)}^\lambda : T_j^\lambda)$$

is invariant to right action by all $U_{p,q}$ $(p > q)$. Since these $U_{p,q}$ generate

$$U^- = U_m^-(K),$$

it follows that $(T_{\ell(\lambda)}^\lambda : T_j^\lambda)$ *is invariant to right action by* U^- (and this for arbitrary $j \in I(n,r)$).

We shall prove in Section (2.6) a converse to this, namely that every right U^--invariant element of $A_r(m \mid n)$ can be written as a linear combination of bideterminants of form $(T^\lambda_{\ell(\lambda)} : T^\lambda_j)$.

(2.3) In this section $\lambda \vdash r$ is fixed. For each pair $p,q \in \mathbf{m}$ $(p > q)$, define a map

$$\Omega = \Omega^\lambda_{p,q} : I(m,r) \to I(m,r) \tag{2.3a}$$

by the rule: if $i \in I(m,r)$, then Ωi is the element of $I(m,r)$ obtained from i by replacing $i_\rho = p$ with q, at *all* places $\rho \in D^\lambda_{p,q}(i)$. In other words, Ωi is the same as $s(D^\lambda_{p,q}(i))$, in the notation of (2.2)(3). Therefore (2.2d) shows that, for all $j \in I(n,r)$:

(2.3b) The coefficient X_d of t^d in $(T^\lambda_i : T^\lambda_j) \circ u_{p,q}(t)$ is $(T^\lambda_{\Omega i} : T^\lambda_j)$. Here $d = |D^\lambda_{p,q}(i)|$.

Remark. Because K is infinite, equations (2.2d) determine $X_0, X_1, \ldots, X_d$ uniquely. Of course any of $X_0, X_1, \ldots, X_d$ may be zero.

Definition (2.3c). For each pair p,q of integers such that $1 \leq q < p \leq m+1$, let $H^\lambda_{p,q}$ be the set of all $i \in I(m,r)$ which satisfy the following conditions:

(i) T^λ_i is standard,

(ii) $\rho \in \mathcal{R}_a(T^\lambda)$, $1 \leq a \leq q-1 \Rightarrow i_\rho = a$, and

(iii) $\rho \in \mathcal{R}_q(T^\lambda) \Rightarrow i_\rho = q$ or $i_\rho \geq p$.

Thus $i \in H^\lambda_{p,q}$ if and only if T^λ_i is standard, and is "canonical in its first q-1 rows" (condition (ii)), while all entries in the qth row of T^λ_i which are not equal to q, are not less than p (condition (iii)).

(2.3d) Note. From the proof of (1.9h), we know that T^λ_i $(i \in I(m,r))$ cannot be standard unless $\tilde{\lambda}_1 \leq m$. Therefore if $\tilde{\lambda}_1 > m$, all the sets $H^\lambda_{p,q}$ are empty. So for the rest of this section *assume that* $\tilde{\lambda}_1 \leq m$.

Example. Take $\lambda = (7, 6, 4, 2, 2)$ and $m \geq 7$. Then

$$T_i^\lambda = \begin{matrix} 1 & 1 & 1 & 1 & 1 & 1 & 1 \\ 2 & 2 & 2 & 2 & 2 & 2 & \\ 3 & 5 & 5 & 6 & & & \\ 4 & 6 & & & & & \\ 5 & 7 & & & & & \end{matrix}$$

belongs to $H_{5,3}^\lambda$, i.e. the element $i \in I(m,21)$ determined by T_i^λ is an element of $H_{5,3}^\lambda$.

It is clear that

$$H_{2,1}^\lambda = \{i \in I(m,r) \mid T_i^\lambda \text{ is standard}\}, \tag{2.3e}$$

and that, for all $p,q \in \mathbf{m}$ $(p > q)$,

$$H_{p,q}^\lambda \supseteq H_{p+1,q}^\lambda; \tag{2.3f}$$

notice that

$$H_{m+1,m-1}^\lambda = H_{m+1,m}^\lambda$$

consists of the single element $\ell(\lambda)$ (see (2.2f)), and that

$$H_{m+1,q}^\lambda = H_{q+2,q+1}^\lambda$$

for $q = 1,2, \ldots, m-1$.

Thus we have a chain of subsets of $H_{2,1}^\lambda$, namely

$$H_{2,1}^\lambda \supseteq H_{3,1}^\lambda \supseteq \ldots \supseteq H_{m+1,1}^\lambda = H_{3,2}^\lambda \supseteq H_{4,2}^\lambda \supseteq \ldots \supseteq H_{m+1,2}^\lambda$$

$$= H_{4,3}^\lambda \supseteq \ldots \supseteq H_{m+1,m}^\lambda = \{\ell(\lambda)\}. \tag{2.3g}$$

Lemma (2.3h). *Let* $\lambda \vdash r$ *and* $p,q \in \mathbf{m}$ $(p > q)$ *be given. If* $i \in H^{\lambda}_{p,q}$, *then* $D^{\lambda}_{p,q}(i)$ *is the set of all* $\rho \in \mathcal{R}_q(T^{\lambda})$ *such that* $i_{\rho} = p$.

Proof. Let $\rho \in \mathcal{R}_q(T^{\lambda})$ be such that $i_{\rho} = p$. Let σ be any element in the column of T^{λ} which contains ρ. If $\sigma \in \mathcal{R}_a(T^{\lambda})$ and $a < q$, then $i_{\sigma} = a$ (condition (2.3c)(ii)). If $\sigma \in \mathcal{R}_a(T^{\lambda})$ and $a > q$, then $i_{\sigma} > i_{\rho} = p > q$ (because T^{λ}_i is standard, (2.3c)(i)). So in either case $i_{\sigma} \neq q$, which shows that $\rho \in D^{\lambda}_{p,q}(i)$.

Conversely suppose $\rho \in D^{\lambda}_{p,q}(i)$; we want to prove that $\rho \in \mathcal{R}_q(T^{\lambda})$. If not, then $\rho \in \mathcal{R}_a(T^{\lambda})$ for some $a > q$ (since $a < q$ would imply $i_{\rho} = a < q$ by (2.3c)(ii)). Let σ be the element of $\mathcal{R}_q(T^{\lambda})$ in the same column as ρ. Because T^{λ}_i is standard, $i_{\rho} > i_{\sigma}$ and hence $i_{\sigma} < p$. Now (2.3c)(iii) implies $i_{\sigma} = q$, hence $\rho \notin D^{\lambda}_{p,q}(i)$, contradiction.

Corollary (2.3i). *If* $i \in H^{\lambda}_{p,q}$ *then* $\Omega^{\lambda}_{p,q}(i) \in H^{\lambda}_{p+1,q}$. *Hence* $\Omega^{\lambda}_{p,q}$ *induces a map*

$$\Omega : H^{\lambda}_{p,q} \to H^{\lambda}_{p+1,q}.$$

Corollary (2.3j). *If* $i, i' \in H^{\lambda}_{p,q}$ *and* $\Omega i = \Omega i'$, *and if also* $|D^{\lambda}_{p,q}(i)| = |D^{\lambda}_{p,q}(i')|$, *then* $i = i'$.

The proofs of (2.3i) and (2.3j) follow easily from Lemma (2.3h), and are left to the reader.

(2.4) Let $\mathcal{P}_m(r)$ be the set of all partitions $\lambda \vdash r$ such that $\tilde{\lambda}_1 \leq m$. A non-empty subset Γ of $\mathcal{P}_m(r) \times I(m,r) \times I(n,r)$ will be called *independent* if for any family of elements $x^{\lambda}_{ij} \in K$,

$$\sum_{(\lambda,i,j)\in\Gamma} x^{\lambda}_{ij}\,(T^{\lambda}_i : T^{\lambda}_j) = 0 \Rightarrow x^{\lambda}_{ij} = 0$$

for all $(\lambda,i,j) \in \Gamma$. In this language, our aim is to prove that the set $\Gamma_{std.}$ of all $(\lambda,i,j) \in \mathcal{P}_m(r) \times I(m,r) \times I(n,r)$ such that T^{λ}_i and T^{λ}_j are both standard, is independent (see (2.1a)). The proof, by induction up the chains of subsets $H^{\lambda}_{p,q}$ of I(m,r), goes back at least to Mead (1972), and in more explicit form (although in a notation different from ours) to de Concini, Eisenbud and Procesi (1980).

For each $\lambda \in \mathbf{P}_m(r)$, we define a subset J_λ of I(n,r). Some of the J_λ may be empty, but we assume that $\bigcup_\lambda J_\lambda$ is not empty. Then for each pair of integers p,q satisfying

$$1 \leq q < p \leq m+1$$

we define the following subset of $\mathbf{P}_m(r) \times I(m,r) \times I(n,r)$:

$$\Gamma_{p,q}(J) = \{(\lambda,i,j) \mid \lambda \in \mathbf{P}_m(r), i \in H^{\lambda}_{p,q}, j \in J_\lambda\}. \tag{1}$$

The essential induction step is provided by the next lemma.

Lemma (2.4a). *Let* J_λ ($\lambda \in \mathbf{P}_m(r)$) *be as above. Let* p,q $\in$ **m** (p > q). *Then if the set* $\Gamma_{p+1,q}(J)$ *is independent, so also is* $\Gamma_{p,q}(J)$.

Proof. Assume that $\Gamma_{p+1,q}(J)$ is independent. To prove that $\Gamma_{p,q}(J)$ is independent, it is enough to prove, for any non-empty subset Γ of $\Gamma_{p,q}(J)$, and any family $\{x^{\lambda}_{ij} \mid (\lambda,i,j) \in \Gamma\}$ of *non-zero* elements of K, that

$$x = \sum_{(\lambda,i,j)\in\Gamma} x^{\lambda}_{ij}\,(T^{\lambda}_i : T^{\lambda}_j) \tag{2}$$

is not zero. Operate on (2) with $u_{p,q}(t)$. By (2.3b)

$$x \circ u_{p,q}(t) = x_0 + tx_1 + \ldots + t^d x_d, \text{ all } t \in K, \tag{3}$$

where $d = \max\{\,|D^{\lambda}_{p,q}(i)| \mid (\lambda,i,j) \in \Gamma\}$ and $x_0, x_1, \ldots, x_d$ are elements of $A_r(m \mid n)$. The coefficient x_d in (3) is given by

$$x_d = \sum_{(\lambda,i,j)\in\Gamma^*} x^{\lambda}_{ij}\,(T^{\lambda}_{\Omega i} : T^{\lambda}_j), \tag{4}$$

where Γ^* is the set of all $(\lambda,i,j) \in \Gamma$, for which $|D^{\lambda}_{p,q}(i)|$ takes the maximum value d. Clearly Γ^* is not empty, and for each $(\lambda,i,j) \in \Gamma^*$, the triple $(\lambda,\Omega i,j)$ lies in $\Gamma_{p+1,q}$, by (2.3i). Moreover $(\lambda,\Omega i,j)$ runs over a certain non-empty subset Γ' of $\Gamma_{p+1,q}(J)$ *without repetitions*, as (λ,i,j) runs over Γ^*.

For suppose

$$(\lambda,i,j),\ (\lambda',i',j') \in \Gamma^*$$

and that

$$(\lambda,\Omega i,j) = (\lambda',\Omega i',j').$$

Then $\lambda = \lambda'$, $j = j'$ and $\Omega i = \Omega i'$. By the definition of Γ^*,

$$|D^{\lambda}_{p,q}(i)| = d = |D^{\lambda}_{p,q}(i')|;$$

hence $i = i'$ by (2.3j).

So the sum in (4) is a non-trivial linear combination of the elements of $\Gamma' \leq \Gamma_{p+1,q}(J)$, and since $\Gamma_{p+1,q}(J)$ is independent, this proves that $x_d \neq 0$. It follows that $x \neq 0$, and Lemma (2.4a) is proved.

The definition (1) of $\Gamma_{p,q}(J)$, together with (2.3g), shows that

$$\Gamma_{2,1}(J) \supseteq \Gamma_{3,1}(J) \supseteq \ldots \supseteq \Gamma_{m+1,m}(J).$$

Applying (2.4a) to successive terms of this sequence, we learn that if

$$\Gamma_{m+1,m}(J) = \{(\lambda,\ell(\lambda),j) \mid \lambda \in \mathbf{P}_m(r),\ j \in J_\lambda\}$$

is independent, then so is $\Gamma_{2,1}(J) = \{(\lambda,i,j) \mid \lambda \in \mathbf{P}_m(r),\ T^{\lambda}_i \text{ standard},\ j \in J_\lambda\}$. It is worth repeating this powerful corollary of (2.4a) in terms of bideterminants.

Corollary (2.4b). *Given a family* J *of subsets* J_λ ($\lambda \in \mathbf{P}_m(R)$) *of* I(m,r) *not all empty, and given that*

$$\{(T^{\lambda}_{\ell(\lambda)} : T^{\lambda}_j) \mid \lambda \in \mathbf{P}_m(r),\ j \in J_\lambda\} \tag{5}$$

is a linearly independent subset of $A_r(m \mid n)$, *then the set*

$$\{(T^{\lambda}_i : T^{\lambda}_j) \mid \lambda \in \mathbf{P}_m(r),\ T^{\lambda}_i \textit{ standard},\ j \in J_\lambda\} \tag{6}$$

is also independent.

Theorem (2.4c). *Let $\mathcal{P}'$ be a non-empty subset of $\mathcal{P}_m(r)$, and for each $\lambda \in \mathcal{P}'$ choose* $j(\lambda) \in I(n,r)$ *such that $T^\lambda_{j(\lambda)}$ is column strict, i.e. the entries in each column of $T^\lambda_{j(\lambda)}$ increase strictly from top to bottom. Then*

$$\{(T^\lambda_i : T^\lambda_{j(\lambda)}) \mid \lambda \in \mathcal{P}',\ T^\lambda_i \textit{ standard}\} \tag{7}$$

is a linearly independent subset of $A_r(m \mid n)$.

Proof. By (2.4b) it is enough to prove that the set

$$\{(T^\lambda_{\ell(\lambda)} : T^\lambda_{j(\lambda)} \mid \lambda \in \mathcal{P}'\} \tag{8}$$

is independent (we take $J_\lambda = \{j(\lambda)\}$ or $\varnothing$, according as λ belongs to $\mathcal{P}'$ or not). Now, for each $\lambda \in \mathcal{P}'$, $(T^\lambda_{\ell(\lambda)} : T^\lambda_{j(\lambda)})$ is non-zero (by 1.7d; both $T^\lambda_{\ell(\lambda)}$ and $T^\lambda_{j(\lambda)}$ are column strict) and has right weight λ (see (2.5), Example 3). Therefore (8) is independent, because its elements lie in distinct right weight spaces of $A_r(m \mid n)$ (see (2.5c)).

Corollary (2.4d). *Let $\lambda \in \mathcal{P}_m(r)$, $j \in I(n,r)$ be such that T^λ_j is column strict (it is not assumed that T^λ_j is standard). Then $\{(T^\lambda_i : T^\lambda_j) \mid i \in I(m,r),\ T^\lambda_i$ standard$\}$ is a linearly independent subset of $A_r(m \mid n)$.*

All the propositions of this chapter have "right-hand" analogues (cf. Section 1.10). To state (and simultaneously prove) the analogue of a proposition P(m,r) concerning $A_r(m \mid n)$, we first interchange m and n, which gives a proposition P(n,r) concerning $A_r(n \mid m)$, and then apply the K-algebra isomorphism

$$\Phi : A(n \mid m) \to A(m \mid n)$$

which takes

$$c_{\nu\mu} \to c_{\mu\nu}\ (\mu \in \mathbf{m},\ \nu \in \mathbf{n}).$$

Using this recipe on Theorem (2.4c), we get its analogue:

Theorem (2.4c'). *Let* $\mathcal{P}'$ *be a non-empty subset of* $\mathcal{P}_n(r)$, *and for each* $\lambda \in \mathcal{P}'$ *choose* $i(\lambda) \in I(m,r)$ *such that* $T^{\lambda}_{i(\lambda)}$ *is column strict. Then*

$$\{(T^{\lambda}_{i(\lambda)} : T^{\lambda}_{j}) \mid \lambda \in \mathcal{P}',\ T^{\lambda}_{j}\ \textit{standard}\} \tag{7'}$$

is a linearly independent subset of $A_r(m \mid n)$.

We can now at last finish the

Proof of Theorem 1. It is enough to prove that

$$\{(T^{\lambda}_{h} : T^{\lambda}_{k}) \mid \lambda \vdash r,\ T^{\lambda}_{h}\ \text{standard},\ T^{\lambda}_{k}\ \text{standard}\} \tag{2.1a}$$

is a linearly independent subset of $A_r(m \mid n)$. Since the fact that T^{λ}_{h} is standard implies that $\lambda_1 \leq m$, i.e. that $\lambda \in \mathcal{P}_m(r)$, we may write (2.1a) as

$$\{(T^{\lambda}_{i} : T^{\lambda}_{j}) \mid \lambda \in \mathcal{P}_m(r),\ T^{\lambda}_{i}\ \text{standard},\ j \in J_{\lambda}\}, \tag{9}$$

where for each λ we put $J_{\lambda} = \{j \in I(n,r) \mid T^{\lambda}_{j}\ \text{standard}\}$. Corollary (2.4b) says that (9) is independent, if

$$\{(T^{\lambda}_{\ell(\lambda)} : T^{\lambda}_{j}) \mid \lambda \in \mathcal{P}_m(r),\ j \in J_{\lambda}\} \tag{10}$$

is independent. Since J_{λ} is empty unless $\tilde{\lambda}_1 \leq n$, i.e. unless $\lambda \in \mathcal{P}_n(r)$, (10) is the same as

$$\{(T^{\lambda}_{\ell(\lambda)} : T^{\lambda}_{j}) \mid \lambda \in \mathcal{P}',\ T^{\lambda}_{j}\ \text{standard}\}, \tag{11}$$

where $\mathcal{P}' = \mathcal{P}_m(r) \cap \mathcal{P}_n(r)$. But (11) is independent by (2.4c').

(2.5) Weights and weight spaces

Before we define the Schur modules $L_{\lambda}(K)$, it will be useful to review the theory of weights for polynomial representations of general linear groups. (For more detail, see [G, Chapter 3].) Let $m,r \in \mathbb{N}$. We make some definitions.

Definition. $\Lambda(m,r) := \{\alpha = (\alpha_1, ..., \alpha_m\} \in N_0^m \mid \alpha_1 + ... + \alpha_m = r\}$.

Definition. An element $i = (i_1, ..., i_r)$ of $I(m,r)$ is said to have *weight* (or *content*) $\alpha \in \Lambda(m,r)$ if:

For all $\mu \in \mathbf{m}$, $\alpha_\mu = |\{\rho \in \mathbf{r} \mid i_\rho = \mu\}|$. (2.5a)

Notation. Write $wt(i) = \alpha$ to indicate that i has weight α.

Example 1. Taking $m = 3$, $r = 5$, we have

(1, 3, 1, 1, 1) has weight (4, 0, 1),

(2, 2, 2, 2, 2) has weight (0, 5, 0),

(3, 2, 1, 3, 2) has weight (1, 2, 2).

Example 2. Let $\lambda \in \mathcal{P}_m(r)$ and let $T^\lambda_{\ell(\lambda)}$ be the canonical λ-tableau (see end of Section (1.9)). Clearly $\ell(\lambda)$ has λ_1 1's, λ_2 2's, and so on. Hence $\ell(\lambda)$ has weight $\lambda = (\lambda_1, ..., \lambda_m)$.

Example 3. Let $i, i' \in I(m,r)$. Then $wt(i) = wt(i')$ if and only if $i \sim i'$ (see (1.5a)). So the P(r)-orbits of I(m,r) correspond bijectively to the elements of $\Lambda(m,r)$.

This purely combinatoric idea of weight comes into the representation theory of $G_m = GL_m(K)$ as follows: each $\alpha \in \Lambda(m,r)$ determines a one-dimensional representation

$$\chi_\alpha : T_m(K) \to K^\times$$

of the group $T_m(K)$ of all diagonal matrices

$$x(t_1, ..., t_m) = \begin{pmatrix} t_1 & & & \\ & \cdot & & \\ & & \cdot & \\ & & & \cdot \\ & & & & t_m \end{pmatrix}$$

of G_m, namely

$$\chi_\alpha(x(t_1, ..., t_m)) = t_1^{\alpha_1} \dots t_m^{\alpha_m}, \text{ all } t_1, ..., t_m \in K^\times. \qquad (2.5a)$$

Now let V be any right KG_m-module. (This means that V is a finite-dimensional space over K, on which G_m acts on the right; V can also be regarded as right module for the (infinite dimensional) group algebra KG_m of G_m over K.)

Definition. An element $v \in V$ is said to be a *weight element of weight* α, or simply to *have weight* α, if

$$vx = \chi_\alpha(x)v, \text{ for all } x \in T_m(K). \tag{2.5b}$$

Definition. The set V^α of all weight elements $v \in V$ of weight α, is called the α*-weight space* of V.

If $V^\alpha \neq \{0\}$, then we say α *is a weight of* V.

A fundamental property of weight spaces, which was already known to Schur (1901), is that if V affords a matrix representation of G_m which is polynomial of degree r, then

$$V = \sum_{\alpha \in \Lambda(m,r)}^{\oplus} V^\alpha. \tag{2.5c}$$

Naturally one can also define weight spaces ${}^\alpha V$ of a left KG_n-module; if V is a (G_n,G_m)-bimodule one must distinguish between "left" and "right" weights.

Example 4. If $i \in I(m,r)$ has weight $\alpha \in \Lambda(m,r)$, and if $j \in I(n,r)$ has weight $\beta \in \Lambda(n,r)$, then one proves (using (2.5b) and (1.4d)) that $c_{i,j} \in A_r(m \mid n)$ is a right weight element of (right) weight α; similarly $c_{i,j}$ is a left weight element of (left) weight β. It follows that the bideterminant

$$(T_i^\lambda : T_j^\lambda) = \sum_{\pi \in C(\lambda)} (-1)^\pi c_{i\pi,j}$$

is also a left (and right) weight element of the (G_n, G_m)-bimodule $A_r(m \mid n)$, and has left (right) weight $\alpha(\beta)$.

(2.6) For our further investigations, we must identify the space A^{U^-} of all right U^-_m-invariant elements of $A = A_r(m \mid n)$.

Definition (2.6a).

$$A^{U^-} := \{x \in A_r(m \mid n) \mid x \circ u = x, \text{ all } u \in U^-_m(K)\}.$$

For each $\lambda \in \mathbf{P}_m(r)$ let J_λ be the set of all $j \in I(n,r)$ such that T_j^λ is standard. This set is non-empty if and only if $\lambda \in \mathbf{P}_m(r) \cap \mathbf{P}_n(r)$ (see (1.9h), and the end of Section (2.4)). Choose p, q such that $1 \leq q < p \leq m+1$ and define, as in Section (2.4),

$$\Gamma_{p,q} := \Gamma_{p,q}(J) := \{(\lambda,i,j) \mid \lambda \in \mathbf{P}_m(r) \cap \mathbf{P}_n(r), i \in H^\lambda_{p,q}, j \in J_\lambda\}. \tag{1}$$

Since T_i^λ is standard for each $i \in H^\lambda_{p,q}$, the set

$$\{(T_i^\lambda : T_j^\lambda) \mid (\lambda,i,j) \in \Gamma_{p,q}\} \tag{2}$$

is a subset of the basis (2.1a) of $A = A_r(m \mid n)$. Let $Y_{p,q}$ denote the K-linear subspace of A spanned by (2). By (2.3e) and (2.3g) we have

$$A = Y_{2,1} \supseteq Y_{3,1} \supseteq \dots \supseteq Y_{m+1,m}, \tag{3}$$

and $Y_{m+1,m}$ has as basis the set

$$\{(T^\lambda_{\ell(\lambda)} : T_j^\lambda) \mid \lambda \in \mathbf{P}_m(r) \cap \mathbf{P}_n(r), T_j^\lambda \text{ standard}\}. \tag{2.6b}$$

We come now to a fundamental theorem, whose first part goes back at least to J Deruyts ([D, §38]). The proof here is based on that in [CEP, Section 3].

Theorem II ([CEP, Theorem 3.3]). *With the notation above:*

(i) $A^{U^-} = Y_{m+1,m}$, *i.e.* A^{U^-} *is the subspace of* $A = A_r(m \mid n)$ *with basis (2.6b).*

(ii) A^{U^-} *is an essential (right)* KU^-*-submodule of* A, *i.e. for every non-zero* $x \in A$, *the set* $x \circ KU^-$ *has non-zero intersection with* A^{U^-}. *Here* KU^- *denotes the group algebra over* K *of the subgroup* $U^- = U^-_m(K)$ *of* $G_m = GL_m(K)$.

Proof. Take $p,q \in \mathbf{m}$ $(p > q)$, and let x be any non-zero element of $Y_{p,q}$. Then x may be taken as the element x in the proof of Lemma (2.4a) (taking the J_λ in (2.4a) to be the set of all $j \in I(n,r)$ with T_j^λ standard). The proof of Lemma (2.4a) shows that there exists an integer $d \geq 0$, and elements $x_0, x_1, \dots, x_d$ in A such that

$$x \circ u_{p,q}(t) = x_0 + tx_1 + \dots + t^d x_d, \text{ all } t \in K; \tag{4}$$

moreover by (2.4a) (4) we see that

$$x_d \in Y_{p+1,q} \quad \text{and} \quad x_d \neq 0. \tag{5}$$

Proof of statement (i) of Theorem II. First, $Y_{m+1,m} \subseteq A^{U^-}$ by (2.6b) and (2.2), Example 2. So we must show that

$$A^{U^-} \subseteq Y_{m+1,m}.$$

If this were false, then by (3) there would exist integers $p,q \in \mathbf{m}$ $(p > q)$ and an element $x \in A^{U^-}$ which lies in $Y_{p,q}$ but not in $Y_{p+1,q}$. In particular $x \neq 0$, so that (4) and (5) hold. But $x \in A^{U^-}$, hence $x \circ u_{p,q}(t) = x$ for all $t \in K$. So we must have $d = 0$ and $x = x_0$; now (5) contradicts the assumption $x \notin Y_{p+1,q}$.

Proof of statement (ii) of Theorem II. If this were false there would exist $0 \neq x \in A$ such that $(x \circ KU^-) \cap A^{U^-} = \{0\}$. Let $Y_{p,q}$ be the last term in (3) which contains such an x. Clearly

$$Y_{p,q} \neq Y_{m+1,m} = A^{U^-}$$

(for $x \in A^{U^-} \Rightarrow x \in (x \circ KU^-) \cap A^{U^-}$). So $p,q \in \mathbf{m}$ $(p > q)$, and our x satisfies (4) and (5).

Since K is infinite, equations (4) show that $x_0, x_1, \ldots, x_d \in x \circ KU^-$. (Replace t in (4) by each of d+1 distinct elements $t_0, t_1, \ldots, t_d$ of K. We get a system of d+1 linear equations for $x_0, x_1, \ldots, x_{d+1}$. These have a solution, since the matrix (t_a^b) $(a,b = 0,1,\ldots,d)$ of this system has non-zero determinant; each x_a is then expressed as a linear combination of

$$x \circ u_{p,q}(t) \ (t = t_0, t_1, \ldots, t_d),$$

hence x_a lies in $x \circ KU^-$.) In particular $0 \neq x_d \in (x \circ KU^-) \cap Y_{p+1,q}$. But our definition of $Y_{p,q}$ implies that $(x \circ KU^-) \cap Y_{p+1,q} = \{0\}$. This contradiction proves (ii), and the proof of Theorem II is complete.

Theorem II can be transformed into the Theorem II' below, by the procedure which transformed Theorem 2.4c into Theorem 2.4c'. Namely we interchange m and n in Theorem II, and then apply the K-algebra isomorphism

$$\Phi : A_r(n \mid m) \to A_r(m \mid n).$$

One checks that

$$\Phi(c \circ u) = u^T \circ \Phi(c)$$

for any $c \in A_r(n \mid m)$, $u \in G_n$; here u^T denotes the transpose of the matrix u. Therefore Φ maps $A_r(n \mid m)^{U^-}$ to ${}^{U^+}A_r(m \mid n)$, where $U^- = U^-_n$ and $U^+ = U^+_n$ are, respectively, the lower and upper unitriangular subgroups of G_n. Remembering also that Φ maps $(T^\lambda_j : T^\lambda_i)$ to $(T^\lambda_i : T^\lambda_j)$ for any $\lambda \vdash r$, $i \in I(m,r)$, $j \in I(n,r)$, we obtain:

Theorem II'. *Let* $A = A_r(m \mid n)$, $U^+ = U^+_n$. *Then*

(i) ${}^{U^+}A$ *is the subspace of* A *with basis*

$$\{(T^\lambda_i : T^\lambda_{\ell(\lambda)}) \mid \lambda \in \mathcal{P}_m(r) \cap \mathcal{P}_n(r),\ T^\lambda_i \textit{ standard}\}. \tag{2.6b'}$$

(ii) ${}^{U^+}A$ *is an essential left* KU^+*-submodule of* A, *i.e. if* $0 \neq x \in A$, *then the set* $KU^+ \circ x$ *has non-zero intersection with* ${}^{U^+}A$.

(2.7) For the rest of these lectures we take m = n. (In the notation of [G, §2.1], $A(n \mid n)$ may be identified with the algebra $A_K(n)$ of polynomial functions on $G_n = GL_n(K)$; then $A_r(n \mid n)$ becomes the space denoted $A_K(n,r)$ in [G].) Keeping the notation of this chapter, $A = A_r(n \mid n)$ is a (G_n, G_n)-bimodule, and

$$U^- = U^-_n \text{ resp. } U^+ = U^+_n$$

are the lower resp. upper unitriangular subgroups of G_n. Recall that $\mathcal{P}_n(r)$ is the set of all partitions λ of r which have not more than n parts, i.e. for which $\tilde{\lambda}_1 \leq n$.

By Theorem II, A^{U^-} has K-basis

$$\{(T^\lambda_{\ell(\lambda)} : T^\lambda_j) \mid \lambda \in \mathcal{P}_n(r), j \in I(n,r), T^\lambda_j \text{ standard}\}.$$

Definition. For each $\lambda \in \mathcal{P}_n(r)$ let $L_\lambda = L_\lambda(K)$ denote the subspace of A^{U^-} having K-basis

$$\{(T^\lambda_{\ell(\lambda)} : T^\lambda_j) \mid j \in I(n,r), T^\lambda_j \text{ standard}\}. \tag{2.7a}$$

$L_\lambda(K)$ is called the *Schur module* for $G_n = GL_n(K)$ corresponding to the partition λ. It coincides with the module $D_{\lambda,K}$ described in [G] (see [G, pp.54-55]).

Theorem III. *Let* $\lambda \in \mathcal{P}_n(r)$. *Then with the notation above*

(i) $L_\lambda = \{c \in A \mid c \circ u = c, c \circ x = \chi_\lambda(x)c,$ *all* $u \in U^-, x \in T_n\}$.

(ii) L_λ *is a left* KG_n-*submodule of* A^{U^-}, *and*

$$A^{U^-} = \sum_{\lambda \in \mathcal{P}_n(r)}^{\oplus} L_\lambda.$$

(iii) *Every non-zero left* KU^+-*submodule of* L_λ *contains the element*

$$(T^\lambda_{\ell(\lambda)} : T^\lambda_{\ell(\lambda)}).$$

Hence L_λ *is indecomposable as left* KU^+-*module, and hence as left* KG_n-*module. The left* KG_n-*socle of* L_λ *is*

$$M_\lambda := KG_n \circ (T^\lambda_{\ell(\lambda)} : T^\lambda_{\ell(\lambda)}),$$

and this is a simple (= irreducible) KG_n-*module. If* char. $K = 0$, *then* $L_\lambda = M_\lambda$ *is a simple (= irreducible)* KG_n-*module.*

(iv) *If* $\alpha \in \Lambda(n,r)$ *is a weight of* L_λ, *i.e. if the left* α-*weight space* ${}^\alpha(L_\lambda)$ *is not zero, then* $\alpha \trianglelefteq \lambda$. *(If* $\alpha, \beta \in \Lambda(n,r)$, *we write* $\alpha \trianglelefteq \beta$ *if*

$$\alpha_1 \leq \beta_1, \alpha_1 + \alpha_2 \leq \beta_1 + \beta_2, \textit{etc.}$$

For properties of this "dominance order" see [JK, §1.4].) *The* λ-*weight space* ${}^\lambda(L_\lambda)$ *of* L_λ *is one-dimensional and is spanned by* $(T^\lambda_{\ell(\lambda)} : T^\lambda_{\ell(\lambda)})$.

(v) *The formal character (see below) of* L_λ *is the Schur function*

$$s_\lambda(X_1, \ldots, X_n).$$

Proof.

(i) The group $U^- = U^-_n(K)$ is normalized by the diagonal subgroup $T_n = T_n(K)$ of G_n, hence A^{U^-}, although it is not a right KG_n-submodule of A, is closed to the right action of T_n. Therefore we may define the right λ-weight space $(A^{U^-})^\lambda$ to be the set of all $c \in A^{U^-}$ such that $c \circ x = \chi_\lambda(x)c$ for all $x \in T_n$. But the right weight of an element $(T^{\mu}_{\ell(\mu)} : T^{\mu}_{j})$ of the basis (2.6b) of A^{U^-} is μ (see (2.5), Example 2). Hence (2.7a) is a basis of $(A^{U^-})^{\lambda}$. This shows that $L_\lambda = (A^{U^-})^{\lambda}$, and (i) follows.

(ii) From the description of L_λ given in (i), it follows that L_λ is closed to the left action of any $g \in G_n$. The second assertion of (ii) is immediate from Theorem II, (i), and the definition of L_λ.

(iii) The bases (2.7a) and (2.6b') of L^λ and ${}^{U^+}A$, respectively, are both subsets of the basis (2.1a) of $A = A_r(n \mid n)$. It follows that the intersection of the bases (2.7a) and (2.6b'), namely the one-element set $\{(T^{\lambda}_{\ell(\lambda)} : T^{\lambda}_{\ell(\lambda)})\}$, is a basis of $L^\lambda \cap {}^{U^+}A$. Theorem II', (ii) now shows that any non-zero left KU^+-submodule of L_λ contains $(T^{\lambda}_{\ell(\lambda)} : T^{\lambda}_{\ell(\lambda)})$; the indecomposability of L_λ as left KU^+- and KG - module follows, and so does the statement in (iii) about the left KG_n - socle M_λ of L_λ. We shall see in Section (3.2) that when char.K is zero, $A = A_r(n \mid n)$ is semisimple as left and right KG_n-module. So in this case L_λ is a semisimple left KG_n-module, from which it follows that $L_\lambda = M_\lambda$ is simple.

(iv) The element $(T^{\lambda}_{\ell(\lambda)} : T^{\lambda}_{j})$ of the basis (2.7a) of L_λ has left weight equal to the weight of j ((2.5), Example 4). Therefore the (left) α-weight space of L_λ has dimension $K_{\lambda\alpha}$ equal to the number of elements in the set

$$\{j \in I(n,r) \mid T^{\lambda}_{j} \text{ standard and } wt(j) = \alpha\}.$$

$K_{\lambda\alpha}$ is the *Kostka number*; see [Ma, pp.56-57], where it is proved that $K_{\lambda\alpha} = 0$ unless $\alpha \trianglelefteq \lambda$.

(v) Suppose that V is a left KG_n-module which affords (relative to some K-basis of V) a polynomial representation R of G_n which is homogeneous of degree r. Then the *formal character* of V is defined to be the following element of the polynomial ring $\mathbf{Z}[X_1, ..., X_n]$:

$$\Phi_V(X_1, ..., X_n) = \sum_{\alpha \in \Lambda(n,r)} (\dim {}^{\alpha}V)\, X_1^{\alpha_1} \dots X_n^{\alpha_n}. \tag{2.7b}$$

It is well known (see for example [G, p.41]) that Φ_V is a symmetric, homogeneous polynomial of degree r in the variables $X_1, ..., X_n$.

The *natural character* of V is defined to be the map $\psi_V : G_n \to K$ given by the formula

$$\psi_V(g) = \mathrm{Tr}\, R(g),\ g \in G_n.$$

These two "characters" of V are connected as follows: if $\zeta_1, ..., \zeta_n$ are the eigenvalues of the matrix R(g) (these lie in some algebraic closure of K), then

$$\psi_V(g) = \Phi_V(\zeta_1, ..., \zeta_n)$$

(see [G, p.42]).

Now take $V = L_\lambda$. By the proof of (iv) above, the α-weight space of L_λ has dimension $K_{\lambda\alpha}$, hence

$$\Phi_{L_\lambda}(X_1, ..., X_n) = \sum_{\alpha \in \Lambda(n,r)} K_{\lambda\alpha}\, X_1^{\alpha_1} \ldots X_n^{\alpha_n}.$$

But the symmetric polynomial on the right is the Schur function $s_\lambda(X_1, ..., X_n)$ (see [Ma, p.42, (5.13)]). This completes the proof of Theorem III.

There is a version of Theorem III which describes the right KG_n-module

$$_\lambda L = \{c \in A \mid u \circ c = c,\ x \circ c = \chi_\lambda(x)c, \text{ all } u \in U_n^+,\ x \in T_n\},$$

which has basis

$$\{(T_i^\lambda : T_{\ell(\lambda)}^\lambda) \mid i \in I(n,r),\ T_i^\lambda \text{ standard}\}. \tag{2.7a'}$$

We leave it to the reader to state and prove the Theorem III' analogous to Theorem III.

(2.8) Let V be finite-dimensional left KG-module. In anticipation of a notation to be introduced in Section (3.1), we say that V belongs to (the category) $M_K(n,r)$, if V affords a matrix representation of G_n which is polynomial and homogeneous of degree r. In particular

$$A = A_r(n \mid n) = A_K(n,r)$$

belongs to $M_K(n,r)$ (see (3.2)), as do also its submodules L_λ and $M_\lambda = \mathrm{soc}\, L_\lambda$ for all $\lambda \in \mathcal{P}_n(r)$.

From Theorem III(iii) we know that each module M_λ is irreducible. We show next that these are, up to isomorphism, the only irreducible objects of $M_K(n,r)$.

Theorem IV. (i) *The set*

$$\{M_\lambda = \text{soc } L_\lambda \mid \lambda \in \mathcal{P}_n(r)\} \tag{2.8a}$$

is a full set of irreducible KG_n*-modules in* $M_K(n,r)$.

(ii) *For each* $\lambda \in \mathcal{P}_n(r)$*, the formal character* $\Phi_{M_\lambda}(X_1, \ldots, X_n)$ *has leading term* $X_1^{\lambda_1} \ldots X_n^{\lambda_n}$ *(the monomials* $X_1^{\alpha_1} \ldots X_n^{\alpha_n}$ *being arranged in the usual lexicographic order).*

Proof (i). First we show that $M_\lambda \not\cong M_\mu$, for any pair of distinct elements $\lambda, \mu \in \mathcal{P}_n(r)$. For if there were a KG_n-isomorphism $f: M_\lambda \to M_\mu$, then $f((T^\lambda_{\ell(\lambda)} : T^\lambda_{\ell(\lambda)}))$ would be a non-zero element of M_μ of (left) weight λ, hence $\lambda \trianglelefteq \mu$ by Theorem III, (iv). In the same way, the existence of the KG_n-isomorphism $f^{-1}: M_\mu \to M_\lambda$ shows that $\mu \trianglelefteq \lambda$. This gives $\lambda = \mu$, contradicting our hypothesis.

It remains to show that *any* irreducible KG_n-module $M \in M_K(n,r)$ is isomorphic to some M_λ. In Section (3.3) it will be shown that M is isomorphic to a submodule of the left KG_n-module $A = A_K(n,r)$; we shall therefore assume that M is a left KG_n-submodule of A. Take any non-zero $x \in M$. By Theorem II(ii), there exists $\xi \in KU^-$ such that

$$0 \neq x \circ \xi \in A^{U^-}.$$

Clearly $M' = M \circ \xi$ is a left KG_n-submodule of A, and is isomorphic to M by the map $a \mapsto a \circ \xi$ $(a \in M)$. But M' is KG_n-generated by $x \circ \xi \in A^{U^-} = \Sigma^\oplus L_\lambda$ (see Theorem II(ii); the sum is over all $\lambda \in \mathcal{P}_n(r)$). Therefore M' lies in

$$\text{soc } A^{U^-} = \Sigma^\oplus M_\lambda$$

(Theorem III(iii)); the M_λ are mutually non-isomorphic irreducible KG_n-modules, and hence $M' = M_\lambda$, for some $\lambda \in \mathcal{P}_n(r)$. Thus $M \cong M_\lambda$, and Theorem IV (i) is proved.

Proof of (ii). It is easy to check that, for any $\alpha \in \Lambda(n,r)$, $\alpha \trianglelefteq \lambda$ implies that $X_1^{\lambda_1} \ldots X_n^{\lambda_n}$ precedes $X_1^{\alpha_1} \ldots X_n^{\alpha_n}$ in the lexicographic order of monomials. It follows from Theorem III (iv) that $\Phi_{L_\lambda}(X_1, \ldots, X_n)$ has leading term

$$X_1^{\lambda_1} \ldots X_n^{\lambda_n}.$$

But M_λ is a KG_n-submodule (hence also a KT_n-submodule) of L_λ, and $\lambda(M_\lambda) = \lambda(L_\lambda)$ because $\lambda(M_\lambda)$ contains $(T^\lambda_{\ell(\lambda)} : T^\lambda_{\ell(\lambda)})$. Theorem IV (ii) follows at once.

(2.8b) Remark. Theorem IV(ii) shows that $\Phi_{M_\lambda(K)}$, $\lambda \in \mathcal{P}_n(r)$, form a **Z**-basis of the set $\mathcal{S}(n,r)$ of all symmetric polynomials in $\mathbf{Z}[X_1, \ldots, X_n]$ which are homogeneous of degree r (see [G, p.47]). If char.K = 0, then $M_\lambda(K) = L_\lambda(K)$ and $\Phi_{M_\lambda(K)}$ is the Schur function $s_\lambda(X_1, \ldots, X_n)$. If char. K = p > 0, no general formula to describe $\Phi_{M_\lambda(K)}$ is yet known.

III. The Schur algebra $S_K(n,r)$

(3.1) This chapter contains an outline of the method whereby Schur reduced the study of those polynomial representations of $GL_n(K)$ which are homogeneous of given degree r, to the representation theory of the finite-dimensional K-algebra (now called) the Schur algebra $S_K(n,r)$. This algebra plays, in the present theory, much the same rôle as that played by the group algebra KG in representation theory of a finite group G.

First we define the category $M_K(n,r)$ as in [G, p.20] or [G', p.127]; its objects are all those left KG_n-modules V (it is always assumed that $\dim_K V < \infty$) whose *coefficient space* cf(V) lies in $A_K(n,r)$. (For the rest of these lectures I use the notations $A_K(n)$, $A_K(n,r)$ in place of $A(n \mid n)$, $A_r(n \mid n)$.) We recall that cf(V) is defined as follows: take any K-basis $\{v_b : b \in B\}$ of V, then define the matrix representation $R : g \mapsto (r_{ab}(g))$ by the equations

$$gv_b = \sum_{a \in B} r_{ab}(g)v_a, \quad g \in G_n, \quad b \in B. \tag{3.1a}$$

The functions r_{ab} lie in the space $\mathcal{F}$ (see (1.1)); cf(V) is defined to be the subspace of $\mathcal{F}$ spanned by the r_{ab}, for all a,b ∈ B; cf(V) is unaffected by a change of basis $\{v_b\}$. The morphisms $f : V \to V'$ between modules V, V' ∈ $M_K(n,r)$, are defined to be the KG_n-homomorphisms. It is clear that any sub- or quotient-module of a KG_n-module V belongs to $M_K(n,r)$ if V does; also any finite direct sum of modules in $M_K(n,r)$ is itself in $M_K(n,r)$.

Let E be an n-dimensional vector space over K, and let $\{e_\nu \mid \nu \in \mathbf{n}\}$ be a basis of E. We make G_n act "naturally" on E by the rule

$$ge_\nu = \sum_{\mu \in \mathbf{n}} g_{\mu,\nu}\, e_\mu, \quad g \in G_n, \quad \nu \in \mathbf{n}.$$

Then G acts on the r-fold tensor product

$$E^r = E \otimes \dots \otimes E \quad (\otimes = \otimes_K)$$

by "diagonal action" : $g(x_1 \otimes \dots \otimes x_n) = gx_1 \otimes \dots \otimes gx_n$, for all $g \in G_n$ and $x_1,\dots,x_n \in E$. Relative to the basis

$$\{e_i = e_{i_1} \otimes \dots \otimes e_{i_r} \mid i \in I(n,r)\}$$

of E^r, this action of G_n is expressed by equations

$$ge_j = \sum_{i \in I} g_{i,j}\, e_i, \quad g \in G_n, \quad j \in I = I(n,r), \tag{3.1b}$$

where $g_{i,j} = g_{i_1 j_1} \dots g_{i_r j_r}$ for all $i,j \in I$. But this shows that the coefficient function corresponding to the pair (i,j) is nothing else than $c_{i,j} = c_{i_1 j_1} \dots c_{i_r j_r}$, see (1.4b). Therefore the coefficient space $cf(E^r) = A_K(n,r)$, which shows that

$$E^r \in M_K(n,r).$$

Let $F : G_n \to GL_I(K)$ be the matrix representation of G_n afforded by equations (3.1b), so that

$$F(g) = (g_{i,j})_{i,j \in I}, \text{ for all } g \in G_n. \tag{3.1c}$$

($M_I(K)$ denotes the K-algebra of all square matrices $(\kappa_{i,j})_{i,j \in I}$ with coefficients $\kappa_{i,j} \in K$; $GL_I(K)$ is the group of all non-singular elements of $M_I(K)$.)

Extending F, by linearity, to the whole of the group-algebra KG_n, we get a K-algebra homomorphism

$$F : KG_n \to M_I(K). \tag{3.1d}$$

Definition. The *Schur algebra* $S_K(n,r)$ is defined to be the image of the map F in (3.1d).

Thus $S_K(n,r)$ is the subalgebra of the matrix algebra $M_I(K)$, consisting of all linear combinations of the matrices $F(g)$, $g \in G_n$. Its importance lies in the fact, first observed by Schur [S], that *every polynomial representation of* G_n, *which is homogeneous of degree* r, *can be factored through* F. In terms of a given KG_n-module V, Schur's discovery can be expressed in the following theorem:

Theorem V. V *belongs to* $M_K(n,r)$ *if and only if* V *is annihilated by the kernel of the map (3.1d).*

The reader is referred to [G', p.128], for a proof of Theorem V. For our present purposes, Theorem V shows that *every* KG_n*-module* V *in* $M_K(n,r)$ *can be regarded as an* $S_K(n,r)$*-module*: the rule is that if $M \in S_K(n,r)$, we take any $x \in XG_n$ such that $F(x) = M$, and then set

$$M.v = xv, \text{ for all } v \in G. \qquad (3.1e)$$

This rule is unambiguous, because if $x' \in KG_n$ also has $F(x') = M$, then

$$x\text{-}x' \in \text{Ker } F,$$

hence (provided $V \in M_K(n,r)$) $xv = x'v$ by Theorem V.

Conversely, *every* $S_K(n,r)$*-module* V *can be regarded as a* KG_n*-module belonging to* $M_K(n,r)$; this time we use (3.1e) to define the action of an element $x \in KG_n$, since M.v is given.

This provides a strong equivalence between $M_K(n,r)$ and the category mod $S_K(n,r)$ of all finite-dimensional, left $S_K(n,r)$-modules: each object $V \in M_K(n,r)$ may be regarded, *via* (3.1e), as object of mod $S_K(n,r)$, and conversely. If W is any K-subspace of V, then clearly W is a KG_n-submodule of V, if and only if it is an $S_K(n,r)$-module of V. If $V,V' \in M_K(n,r)$, then a K-map (= linear map)

$$f : V \to V'$$

is a KG_n-map, if and only if it is an $S_K(n,r)$-map. It follows that a KG_n-module $V \in M_K(n,r)$ is irreducible if and only if V is irreducible as $S_K(n,r)$-module, and that KG_n-modules V, $V' \in M_K(n,r)$ are isomorphic, if and only if they are isomorphic as $S_K(n,r)$-modules.

Everything we have said about the categories $M_K(n,r)$ and mod $S_K(n,r)$, has analogues for the appropriate categories $M'_K(n,r)$ and mod'$S_K(n,r)$ of right KG_n- and right $S_K(n,r)$-modules.

Proposition (3.1f). *The* (KG_n, KG_n)*-bimodule* $A_K(n,r) = A_r(n \mid n)$ *belongs to* $M_K(n,r)$ *(as left* KG_n*-module) and to* $M'_K(n,r)$ *(as right* KG_n*-module). Therefore* $A_K(n,r)$ *may be regarded as an* $(S_K(n,r), S_K(n,r))$*-bimodule. Each element* $M \in S_K(n,r)$ *acts on* $A_K(n,r)$ *by the rules:*

$$M \circ c_{i,j} = \sum_{t \in I} M_{t,j}\, c_{i,t} \tag{3.1g}$$

and

$$c_{i,j} \circ M = \sum_{s \in I} M_{i,s}\, c_{s,j}, \tag{3.1h}$$

for all $i,j \in I$.

Proof. Let $i,j \in I$. Then equations (1.4c) and (3.1c) show that the equation

$$x \circ c_{i,j} = \sum_{t \in I} F(x)_{t,j}\, c_{i,t} \tag{3.1i}$$

holds when x is any element g of G_n. Hence (3.1i) holds for all $x \in KG_n$. In particular, this shows that any $x \in \operatorname{Ker} F$ annihilates $A_K(n,r)$, and so, by Theorem V, $A_K(n,r)$ (regarded as left KG_n-module) belongs to $M_K(n,r)$. Now (3.1g) follows from (3.1i), since by (3.1e), $M \circ c_{i,j}$ is defined to be $x \circ c_{i,j}$, where x is any element of KG_n such that $F(x) = M$. The "right-hand" statements in Proposition (3.1f) are proved similarly.

Corollary (3.1j). L_λ, M_λ *are left* KG_n*-submodules of* $A_K(n,r)$ *(see Theorem III, Section (2.7)), hence they are left* $S_K(n,r)$*-submodules of* $A_K(n,r)$*. By Theorem IV (Section (2.8)) the set*

$$\{M_\lambda = \operatorname{soc} L_\lambda \mid \lambda \in \mathcal{P}_n(r)\}$$

is a full set of irreducible left $S_K(n,r)$*-modules. There hold corresponding statements about the right* $S_K(n,r)$*-modules* ${}_\lambda L$, ${}_\lambda M = \operatorname{soc} {}_\lambda L$.

(3.2) From (3.1c) it is clear that, for any $g \in G_n$, the coefficients of the matrix $M = F(g)$ satisfy

$$(i,j) \sim (i',j') \Rightarrow M_{i,j} = M_{i',j'}, \text{ for all } i,j,i',j' \in I = I(n,r). \tag{3.2a}$$

Since $M_K(n,r)$ is the linear span of the matrices $F(g)$, $g \in G_n$, it follows that (3.2a) holds for all $M \in S_K(n,r)$. But the converse holds as well.

Theorem VI (Schur, Weyl). $S_K(n,r)$ *is the set of all matrices* $M \in M_I(K)$ *which satisfy (3.2a).*

A proof of Theorem VI can be found in Weyl's book [W, pp.98, 130].

Let i,j be any elements of $I = I(n,r)$, and let $\xi_{i,j} \in M_I(K)$ be the matrix whose (h,k)-coefficient is 1 or 0, according as $(h,k) \sim (i,j)$ or not. By this definition:

(3.2b) $\xi_{i,j} = \xi_{i',j'}$ *if and only if* $(i,j) \sim (i',j')$, *and if* Ω *is a set of representatives of the* P(r)*-orbits* ($\sim$*-classes*) *of* $I \times I$, *then by Theorem VI*

$$\{\xi_{i,j} \mid (i,j) \in \Omega\} \tag{3.2c}$$

is a K*-basis of* $S_K(n,r)$.

(3.2b) may be compared with (1.5a), which says that the monomials $c_{i,j}$, $c_{i',j'}$ are equal if and only if $(i,j) \sim (i',j')$. It is clear that $\{c_{i,j} \mid (i,j) \in \Omega\}$ is a K-basis of $A_K(n,r)$ ($= A_r(n \mid n)$), hence

$$\dim S_K(n,r) = \dim A_K(n,r) = \binom{n^2 + r - 1}{r}. \tag{3.2d}$$

In the next section we shall give a pairing between the spaces $S_K(n,r)$ and $A_K(n,r)$ which will show that these are "dual" to each other as $(S_K(n,r), S_K(n,r))$-bimodules.

Special case char.K = 0. The following fundamental theorem is due to Schur ([S, §30]).

Theorem VII. *If* K *is any field of characteristic zero, then the algebra* $S_K(n,r)$ *is semisimple. It follows that every* $S_K(n,r)$*-module is completely reducible (= semisimple), and hence that every module (or representation) in the category* $M_K(n,r)$ *is completely reducible.*

Proof. It will be enough to prove that the algebra $S_K(n,r)$ is semisimple; the statement about complete reducibility then follows by a standard theorem on algebras.

Write $S = S_K(n,r)$. We define a bilinear form $f : S \times S \to K$ by the formula

$$f(\xi,\eta) = \mathrm{Tr}(\xi\eta),\ \xi, \eta \in S.$$

It is easy to calculate the trace of any product $\xi_{i,j}\, \xi_{h,k}$ of elements $\xi_{i,j}$, $\xi_{h,k}$ in our basis (3.2c), and we find that

$$f(\xi_{i,j}, \xi_{h,k}) = [P : P_{i,j}].1_K \text{ or zero,} \tag{3.2e}$$

according as $\xi_{i,j} = \xi_{h,k}$ or $\xi_{i,j} \neq \xi_{h,k}$. Here $[P : P_{i,j}]$ is the index in $P = P(r)$ of the stabilizer

$$P_{i,j} = \{\pi \in P \mid i\pi = i, j\pi = j\}$$

of the pair (i,j). Since char.K = 0, $[P : P_{i,j}].1_K \neq 0$, and it follows at once from (3.2e) that the bilinear form f is non-singular. Now let ξ be any element of the radical rad S of S. Then the matrix $\xi\eta$ is nilpotent, hence has trace zero, for all $\eta \in S$. But this means that $f(\xi,\eta) = 0$ for all $\eta \in S$, and the non-singularity of f then implies $\xi = 0$. Therefore rad S = {0}, i.e. S is semisimple.

Remark. The proof shows that $S_K(n,r)$ is semisimple even when p = char.K is finite, provided p > r. For in this case p does not divide |P| = r!, and hence the bilinear form f is non-singular by (3.2e).

(3.3) Contravariant duality. Assume that $S = S_K(n,r)$, where K is an infinite field of arbitrary characteristic. Let J denote the linear operator on S which replaces each element of S by its matrix transpose. It is clear that $J : S \to S$ is an involutory anti-automorphism of the algebra S, i.e. that

$$J(1_S) = 1_S,\ J(J(x)) = x \text{ and } J(xy) = J(y)J(x), \tag{3.3a}$$

for all $x,y \in S$. Moreover $J(\xi_{i,j}) = \xi_{j,i}$ for all $i,j \in I = I(n,r)$, and if $g \in G_n$, then (3.1c) gives at once

$$J(F(g)) = F(g^{tr}), \tag{3.3b}$$

where g^{tr} denotes the transpose of the matrix g.

Given a module $V \in$ mod S, define $V^0 \in$ mod S to be the dual K-space $Hom_K(V,K)$, with S acting on the left by the rule

$$(\xi\lambda)(v) = \lambda(J(\xi)v), \text{ for } \xi \in S, \lambda \in V^0, v \in V. \tag{3.3c}$$

If V is regarded as a left KG_n-module according to the formula (3.1e), then by (3.3b) and (3.3c) V^0 becomes a left KG_n-module, with G_n acting as follows:

$$(g\lambda)(v) = \lambda(g^{tr}v), \text{ for } g \in G_n, \lambda \in V^0, v \in V. \tag{3.3d}$$

For comments on this definition, see [G, p.32]. Notice that the natural isomorphism $V \to (V^0)^0$ (taking each $v \in V$ to the element

$$\{\lambda \mapsto \lambda(v) \mid \lambda \in V^0\}$$

of $(V^0)^0)$ is an S-map (hence also a KG_n-map), so that $(V^0)^0 \cong V$.

Definition. Let $V, W \in \text{mod } S$. A bilinear form $\phi : V \times W \to K$ is called *contravariant* if

$$\phi(\xi v, w) = \phi(v, J(\xi)w) \text{ for all } \xi \in S, v \in V, w \in W. \tag{3.3e}$$

A bilinear form $\phi : V \times W \to K$ defines and is defined by the K-linear map

$$f : V \to W^0 = \text{Hom}_K(W,K)$$

for which

$$f(v)(w) = \phi(v,w), \text{ all } v \in V, w \in W. \tag{3.3f}$$

It is easily verified that f is bijective if and only if the form ϕ is non-singular, i.e. if ϕ is a "pairing" of V and W, and that f is an S-map if and only if ϕ is contravariant in the sense just defined.

Lemma (3.3g). *Let* $V, W \in \text{mod } S$. *Then:*

(i) $V \cong W^0$ *if and only if there exists a contravariant pairing* $\phi : V \times W \to K$.

(ii) *If such a contravariant pairing* ϕ *exists, there is a bijective correspondence between the sets* Sub(V) *and* Sub(W) *of all S-submodules of* V *and* W, *respectively; to each* $V_1 \in \text{Sub}(V)$ *corresponds its "orthogonal"*

$$V_1^{\perp} = \{w \in W \mid \phi(v_1, w) = 0, \forall\, v_1 \in V_1\},$$

and to $W_1 \in \text{Sub}(W)$ *corresponds*

$$W_1^{\perp} = \{v \in V \mid \phi(v, w_1) = 0, \forall\, w_1 \in W_1\}.$$

This correspondence reverses inclusions, e.g.

$$V_1 \leq V_2 \Rightarrow V_1^{\perp} \geq V_2^{\perp}$$

for all $V_1, V_2 \in \text{Sub}(V)$.

(iii) *If* W *is irreducible, so is* W^0.

Proof. (i) follows from the remarks made below (3.3f). (ii) is straightforward linear algebra (see [CR, pp.318-410]), and (iii) follows at once from (ii) (taking $V = W^0$).

Finally there are definitions and propositions regarding contravariant right S-modules, exactly analogous to those which we have given for left S-modules; we shall not write these out.

Theorem VIII. *Define the bilinear form* $\phi : A_K(n,r) \times S_K(n,r) \to K$ *by*

$$\phi(c_{i,j}, \xi_{h,k}) = \begin{Bmatrix} 1 & \textit{if} \\ 0 & \textit{if not} \end{Bmatrix} (i,j) \sim (k,h), \tag{3.3h}$$

for all $i,j,h,k \in I = I(n,r)$. *Then* ϕ *is non-singular, and is contravariant when* $A_K(n,r)$, $S_K(n,r)$ *are both regarded as left* $S_K(n,r)$*-modules, and also when both are regarded as right* $S_k(n,r)$*-modules. Therefore the map*

$$f : A_K(n,r) \to S_K(n,r)^0$$

defined by ϕ *(see (3.3f)), is an* $(S_K(n,r), S_K(n,r))$*-bimodule isomorphism.*

Proof. That ϕ is non-singular is immediate from (3.3h) and the fact that

$$\{c_{i,j} \mid (i,j) \in \Omega\} \text{ and } \{\xi_{h,k} \mid (k,h) \in \Omega\}$$

are bases of $A_K(n,r)$ and $S_K(n,r)$, respectively (see Section (3.2)). To show that ϕ is contravariant in both the 'left' and 'right' senses, it is enough to verify that

$$\phi(M \circ c, \xi) = \phi(c, J(M)\xi) = \phi(c \circ J(\xi), J(M)), \tag{3.3i}$$

for all $M, \xi \in S_K(n,r)$ and all $c \in A_K(n,r)$. Because ϕ is bilinear, it will be enough to prove (3.3i) for $c = c_{i,j}$, $\xi = \xi_{h,k}$. Routine calculations using (3.1g), (3.1h) show that in this case the three expressions in (3.3i) are all equal to $\sum_s M_{s,j}$, sum over all $s \in I$ such that $(s,i) \sim (h,k)$. This completes the proof of Theorem VIII.

At this point we may prove that any irreducible KG_n-module $M \in M_K(n,r)$ is isomorphic to a submodule of the (left) $S_K(n,r)$-module $A_K(n,r)$; this fills the gap left in the proof of Theorem IV(i) (Section (2.8)). Regard M as S-module, then its dual M^0 is irreducible by (3.3g)(iii). Take any non-zero $m^0 \in M^0$, and define the map

$$F : S_K(n,r) \to M^0$$

by $F(\xi) = \xi m^0$, for all $\xi \in S = S_k(n,r)$. It is clear that F is an S-epimorphism.

Hence the dual map

$$F^0 : (M^0)^0 \to S_K(n,r)^0$$

is an S-monomorphism. Since $(M^0)^0 \cong M$ and $S_K(n,r)^0 \cong A_K(n,r)$, our assertion follows.

Weights of V^0. Suppose $V \in \text{mod } S$ and that $\alpha \in \Lambda(n,r)$ (see Section (2.5)). It is a very useful fact that the α-weight spaces of V and V^0 are always isomorphic (as K-spaces), see [G, p.40]. Consequently the formal characters Φ_V and Φ_{V^0} coincide (see (2.7b)). Moreover every irreducible S-module M_λ (see (2.8a)) is self-dual. For M_λ^0 is irreducible by (3.3g)(iii), and so $M_\lambda^0 \cong M_\nu$ for some $\nu \in \mathcal{P}_n(r)$, by Theorem IV(i). But since $\Phi_{M_\lambda^0} = \Phi_{M_\lambda}$, we have $\Phi_{M_\lambda} = \Phi_{M_\nu}$, hence $\lambda = \nu$ by Theorem IV(ii).

Weyl modules. If $\lambda \in \mathcal{P}_n(r)$, we may define the *Weyl module* $V_\lambda(K)$ to be the dual of the corresponding Schur module: $V_\lambda(K) := L_\lambda(K)^0$. Thus $V_\lambda(K)$ is a left $K.GL_n(K)$-module in the category $M_K(n,r)$, having the same formal character $s_\lambda(X_1, ..., X_n)$ as $L_\lambda(K)$. $V_\lambda(K)$ is isomorphic to the module $V_{\lambda,K}$ defined in [G, p.65]. In general $V_\lambda(K)$ is not isomorphic to $L_\lambda(K)$, but there is an inclusion-reversing bijection from $\text{Sub}(L_\lambda(K))$ to $\text{Sub}(V_\lambda(K))$ (see (3.3g)(ii)). Since $L_\lambda(K)$ has a unique minimal submodule $M_\lambda(K)$, it follows that $V_\lambda(K)$ has a unique maximal submodule $V_\lambda(K)^{max}$; the irreducible KG_n-module $V_\lambda/V_\lambda(K)^{max}$ is easily shown to be isomorphic to $M_\lambda(K)^0 \cong M_\lambda(K)$. For other properties of Weyl modules, see [G, Chapter 5].

(3.4) In this section we fix $\lambda \in \mathcal{P}_n(r)$ and, for the moment, take $K = \mathbf{Q}$. The Schur module $L_\lambda(\mathbf{Q})$ is irreducible (Theorem III(iii)) and has $\mathbf{Q}$-basis

$$\left\{\left(T^\lambda_{\ell(\lambda)} : T^\lambda_j\right) \mid j \in I(\lambda)\right\}, \tag{3.4a}$$

where $I(\lambda) = \{j \in I(n,r) \mid T^\lambda_j \text{ is standard}\}$, see (2.7a).

The matrix representation $g \to R_\lambda(g) = (r_{ab}(g))$ of $G_n = GL_n(\mathbf{Q})$ afforded by (3.4a) is given by the equations

$$g \circ \left(T^{\lambda}_{\ell(\lambda)} : T^{\lambda}_{b}\right) = \sum_{a \in I(\lambda)} r_{ab}(g) \left(T^{\lambda}_{\ell(\lambda)} : T^{\lambda}_{a}\right), \tag{3.4b}$$

for all $g \in G_n$, $b \in I(\lambda)$. The coefficients $r_{ab}(g)$ can be calculated, at least in principle, as follows: first we have by (1.9a)

$$g \circ \left(T^{\lambda}_{\ell(\lambda)} : T^{\lambda}_{b}\right) = \sum_{t \in I(n,r)} g_{t,b} \left(T^{\lambda}_{\ell(\lambda)} : T^{\lambda}_{t}\right). \tag{3.4c}$$

Next we use the "straightening process" to express each bideterminant $(T^{\lambda}_{\ell(\lambda)} : T^{\lambda}_{t})$ in terms of standard ones: we found in Lemma (1.10e') that

$$\left(T^{\lambda}_{\ell(\lambda)} : T^{\lambda}_{t}\right) = \sum_{a \in I(\lambda)} w'_{a,t}\left(T^{\lambda}_{\ell(\lambda)} : T^{\lambda}_{a}\right) + F'_{\ell(\lambda)}, \tag{3.4d}$$

where the $w'_{a,t} \in \mathbf{Z}$ and

$$F'_{\ell(\lambda)} \in \sum_{\mu < \lambda} \sum_{i,j \in I(n,r)} \mathbf{Z} \cdot \left(T^{\mu}_{i} : T^{\mu}_{j}\right). \tag{3.4e}$$

Moreover, by using Theorem (1.9e), we may assume that the sum in (3.4e) is restricted to those i,j for which T^{μ}_{i} and T^{μ}_{j} are standard. In fact $F'_{\ell(\lambda)} = 0$. For the right weight of $(T^{\mu}_{i} : T^{\mu}_{j})$ is the same as the weight of i (Section (2.5), Example 4). But if $\mu < \lambda$, then the Kostka number $K_{\mu\lambda}$ is zero (see the proof of Theorem III(iv), Section (2.7), and notice that $\lambda \ntrianglelefteq \mu$), so that no *standard* bideterminant

$$(T^{\mu}_{i} : T^{\mu}_{j})$$

has right weight λ. But all the terms $(T^{\lambda}_{\ell(\lambda)} : T^{\lambda}_{t})$ and $(T^{\lambda}_{\ell(\lambda)} : T^{\lambda}_{a})$ in (3.4d) have right weight λ, and hence $F'_{\ell(\lambda)} = 0$.

Combining (3.4c) and (3.4d), and comparing the result with (3.4b), we arrive at the formula

$$r_{ab}(g) = \sum_{t \in I(n,r)} w'_{a,t}\, g_{t,b}, \tag{3.4f}$$

for all $a,b \in I(\lambda)$ and all $g \in G_n$; equivalently

$$r_{ab} = \sum_{t \in I(n,r)} w'_{a,t}\ c_{t,b}, \text{ all } a,b \in I(\lambda). \tag{3.4g}$$

This shows that our matrix representation R_λ is an *integral form* in the sense described in the Introduction. Moreover if, in the above calculations, we replace **Q** by an arbitrary infinite field K, then the coefficients $r_{ab}(g)$ of the matrix representation $R_\lambda(K)$ which takes the place of R_λ are given by precisely the same formula (3.4f), since the coefficients $w'_{a,t}$ are independent of K (see Lemmas (1.10e), (1.10e')). It is then easy to show that the family $\{L_\lambda(K)\}$ is *defined over* **Z** in the sense explained in [G, p.30]. It follows that the same is true of the family $\{V_\lambda(K)\}$ of Weyl modules [G, Example 2, p.33]).

Example. Take $n = r = 2$ and $\lambda = (2,0)$. The basis (3.4a) (= (2.7a)) of the Schur module $L_{(2,0)}$ is $\left\{c_{11}^2, c_{11}c_{12}, c_{12}^2\right\}$. It is easy to calculate the matrix representation of G_n afforded by this basis from the formulae

$$g \circ c_{11} = g_{11}c_{11} + g_{21}c_{12},\quad g \circ c_{12} = g_{12}c_{11} + g_{22}c_{12}; \tag{3.4h}$$

no "straightening" is needed! We get the representation $R_{(2,0)}$ of the Introduction. The other representation $R'_{(2,0)}$ can be described by the formula

$$R'_{(2,0)}(g) = R_{(2,0)}(g^{tr})^{tr}.$$

This is the representation afforded by the Weyl module $V_{(2,0)} = L^0_{(2,0)}$, using the basis dual to the basis used to get $R_{(2,0)}$.

If char.K $\neq$ 2 then, by Theorem VII (Section (3.2)) and the subsequent remark, the Schur algebra $S_K(2,2)$ is semisimple. Hence $L_{(2,0)} = M_{(2,0)}$ is simple. Also

$$V_{(2,0)} = L^0_{(2,0)} \cong M^0_{(2,0)} \cong V_{(2,0)}$$

in this case.

If char.K = 2, we calculate the simple module $M_{(2,0)}$ from the definition

$$M_\lambda = KG_n \circ (T^\lambda_{\ell(\lambda)} : T^\lambda_{\ell(\lambda)})$$

(see Theorem III(iii), Section (2.7)). In the present case $\lambda = (2,0)$,

$$\ell(\lambda) = (1,1) \in I(2,2) \text{ and } (T^{\lambda}_{\ell(\lambda)} : T^{\lambda}_{\ell(\lambda)}) = c^2_{11}.$$

Using (3.4h) we find that $M_{(2,0)}$ has basis $\{c^2_{11}, c^2_{12}\}$. Relative to the basis $\{c^2_{11}, c^2_{12}, c_{11}\, c_{12}\}$, $L_{(2,0)}$ affords the reduced matrix representation

$$g \mapsto \begin{bmatrix} g^2_{11} & g^2_{12} & g_{11}g_{12} \\ g^2_{21} & g^2_{22} & g_{21}g_{22} \\ 0 & 0 & g_{11}g_{22}+g_{21}g_{12} \end{bmatrix}.$$

The top left 2×2 submatrix gives the irreducible representation corresponding to $M_{(2,0)}$, and the bottom right 1×1 submatrix corresponds to $L_{(2,1)} = M_{(1,1)}$ (notice $g_{11}\, g_{22} + g_{21}\, g_{12} = g_{11}\, g_{22} - g_{21}\, g_{12} = \det(g)$). We leave it to the reader to prove that $L_{(2,0)} \not\cong V_{(2,0)}$ in case char.K = 2.

Bibliography

[AB] K Akin & D Buchsbaum, Characteristic-free representation theory of the general linear group, I, *Adv. in Math.* **58** (1985), 149-200.

[CL] R Carter & G Lusztig, On the modular representations of the general linear and symmetric groups, *Math. Z.* **136** (1974), 193-242.

[CEP] C de Concini, D Eisenbud & C Procesi, Young diagrams and determinantal varieties, *Invent. Math.* **56** (1980), 129-165.

[Cl] M Clausen, Letter place algebras and a characteristic-free approach to the representation theory of the general linear and symmetric groups, I, *Adv. in Math.* **33** (1979), 161-191.

[Cl'] M Clausen, Kombinatorische Strukturen in Polynomringen, Actes 14ᵉ Séminaire Lotharingien, Publ. I.R.M.A. Strasbourg 323/S-14, 43-46.

[CR] C W Curtis & I Reiner, *Representation theory of finite groups and associative algebras* (J Wiley, New York 1962).

[D] J Deruyts, Essai d'une théorie générale des formes algébriques, *Mém. Soc. Roy. Sci. Liège* **17** (1890), 156.

[DKR] J Désarménien, J P S Kung & G-C Rota, Invariant theory, Young bitableaux and combinatorics, *Adv. in Math.* **27** (1978), 63-92.

[DRS] P Doubilet, G-C Rota & J Stein, On the foundations of combinatorial theory: IX; Combinatorial methods in invariant theory, *Stud. Appl. Math.* **53** (1974), 185-216.

[F] G Frobenius, Über Gruppencharaktere, *Sitzber. Preuss. Akad. Wiss. Berlin* (1896), 985-1021.

[F'] G Frobenius, Über die Charaktere der symmetrischen Gruppe, *Sitzber. Preuss. Akad. Wiss. Berlin* (1900), 516-534.

[G] J A Green, *Polynomial representations of* GL_n (Lecture Notes in Math. **830**, Springer Verlag, Berlin, Heidelberg, New York, 1980).

[G'] J A Green, *Polynomial representations of* GL_n (Lecture Notes in Math. **848**, Springer Verlag, Berlin, Heidelberg, New York, 1980), 124-140.

[J] G D James, *The representation theory of the symmetric groups* (Lecture Notes in Math. **682**, Springer Verlag, Berlin, Heidelberg, New York, 1978).

[JK] G D James & A Kerber, *The representation theory of the symmetric group* (Addison-Wesley, London, Amsterdam, Don Mills, Sydney, Tokyo, 1981).

[M] D G Mead, Determinantal ideals, identities and the Wronskian, *Pacific J. Math.* **42** (1972), 165-175.

[Ma] I G Macdonald, *Symmetric functions and Hall polynomials* (Oxford University Press, Oxford, 1979).

[S] I Schur, Über eine Klasse von Matrizen, die sich einer gegebenen Matrix zuordnen lassen; in *I. Schur, Gesammelte Abhandlungen I* (Springer, Berlin, 1983), 1-70.

[W] H Weyl, *The classical groups* (Princeton University Press, Princeton N.J., 1946).

ON COXETER'S GROUPS $G^{P,Q,R}$

LARRY C GROVE

University of Arizona, Tucson AZ 85721, USA

JANET M MCSHANE

Northern Arizona University, Flagstaff AZ 86011, USA

Coxeter [2] defined the group $G^{p,q,r}$ by the presentation

$$< A,B,C \mid A^p = B^q = C^r = (AB)^2 = (BC)^2 = (CA)^2 = (ABC)^2 = 1>.$$

He interpreted the group as the group of automorphisms of a regular map $\{p,q\}_r$, with presentation

$$<R_1,R_2,R_3 \mid R_1^2 = R_2^2 = R_3^2 = (R_1R_2)^p = (R_2R_3)^q = (R_3R_1)^2 = (R_1R_2R_3)^r = 1>$$

(see [5, p. 111]; think of the generators as reflections in the sides of a fundamental region). The two presentations are equivalent via

$$A = R_1R_2,\ B = R_2R_3,\ C = R_3R_2R_1;$$

$$R_1 = BC,\ R_2 = BCA,\ R_3 = CA.$$

Coxeter identified the $G^{p,q,r}$ with familiar groups for a number of small values of the parameters and established some criteria for the groups to be finite or infinite. In particular he showed that $G^{3,7,r}$ is finite for $r \le 15$. Leech and Mennicke ([8], see also [3]) showed that $G^{3,7,16}$ is finite. It has been shown (see [4, p.12] and [7, p.52]) that $G^{3,7,17}$ is finite (in fact trivial), and Coxeter has conjectured that $G^{3,7,r}$ is infinite for all $r \ge 18$. Sims established that $G^{3,7,18}$ is infinite ([9], see also [6]).

We first investigated $G^{3,7,20}$, using Cayley [1], in response to a question raised by Steve Wilson, and showed it to be infinite. Unaware at the time of [9] and [6], we re-established the fact that $G^{3,7,18}$ is infinite, and subsequently have also established that $G^{3,7,r}$ is infinite for $r = 24, 26, 27$ and 32 (hence, of course, also for any multiples of these values).

Our procedure has been straightforward. We used a short Cayley program to determine low- index subgroups, to abelianize them, and to calculate the resulting invariants via Smith normal form. In all cases of success we have found infinite subgroups of index less than 100. Other values of r (up to 35) have either given us no proper subgroups of index 125 or less, or else the subgroups are all finite when abelianized. We have found that increasing the index from 100 to 125 has greatly increased the required CPU time (we used a Sun-3 workstation at the University of Arizona), as has increasing the size of r.

The lowest indices and resulting invariants are tabulated below; a 0 in the list of invariants indicates the presence of an infinite cyclic summand in the abelianized subgroup. Generators for the subgroups are available on request.

r	index	invariants
18	56	(2,0)
20	84	(2,10,0)
24	84	(2,2,0)
26	42	(2,2,0)
27	84	(2,0)
32	84	(2,2,0)

To the best of our knowledge these are the only values of r for which $G^{3,7,r}$ is known to be infinite. It is stated in the Cayley manual ([1, pp.5-11]) that $G^{3,7,19}$ is infinite, but we have thus far been unable to verify that or to ascertain a source for the statement.

References

1. J J Cannon, *A language for group theory* (University of Sydney 1982), preprint.
2. H S M Coxeter, The abstract groups $G^{m,n,p}$, *Trans. Amer. Math. Soc.* **45** (1939), 73-150.
3. H S M Coxeter, The abstract group $G^{3,7,16}$, *Proc. Edinburgh Math. Soc.* (II) **13** (1962), 47-61 and 189.
4. H S M Coxeter, *Twisted Honeycombs* (Regional Conference Series in Math. No. 4, Amer. Math. Soc., Providence, R.I. 1970).

5. H S M Coxeter and W O J Moser, *Generators and relations for discrete groups, 4th ed.* (Springer-Verlag, Berlin 1980).

6. J Leech, Note on the abstract group (2,3,7;9), *Proc. Cambridge Philos. Soc.* **62** (1966), 7-10.

7. J Leech, Computer proof of relations in groups, in *Topics in group theory and computation* (ed. M P J Curran, Academic Press, London 1977), 38-61.

8. J Leech and J Mennicke, Note on a conjecture of Coxeter, *Proc. Glasgow Math. Assoc.* **5** (1969), 25-29.

9. C C Sims, On the group (2,3,7;9), *Notices Amer. Math. Soc.* **11** (1964), 687-688.

INTEGRAL DIMENSION SUBGROUPS

NARAIN GUPTA

University of Manitoba, Winnipeg, R3T 2N2, Canada

1. Introduction

Let G be a group and let **Z**G denote its integral group ring. Consider the ring homomorphism $\varepsilon:\mathbf{Z}G \to \mathbf{Z}$ given by $\Sigma\, n_i g_i \to \Sigma\, n_i$. The kernel of ε is known as the *augmentation ideal* of **Z**G and we denote it by $\Delta(G)$ (other names in the literature for this ideal are the *Magnus ideal* and the *fundamental ideal* of **Z**G). Alternately,

$$\Delta(G) = \mathbf{Z}G\,(G\text{-}1) = (G\text{-}1)\mathbf{Z}G$$

is the ideal of **Z**G generated by all elements g-1, $g \in G$. For each $n \geq 1$, let $\Delta^n(G)$ be the ideal generated by the products of the form $(g_1\text{-}1)...(g_n\text{-}1)$, $g_i \in G$ and let

$$D_n(G) = G \cap (1+\Delta^n(G))$$

be the subset of G consisting of all elements $g \in G$ with the property that $g-1 \in \Delta^n(G)$. Since for $g,h \in G$,

$$gh\text{-}1 = (g\text{-}1) + g(h\text{-}1),$$

$$(g^{-1} - 1) = - g^{-1}(g\text{-}1)$$

and

$$(h^{-1}gh-1) = h^{-1}(g-1)h,$$

it follows that $D_n(G)$ is a normal subgroup of G known as the *n-th dimension subgroup* of G. Let $\gamma_n(G)$, $n \geq 1$, denote the subgroup of G generated by all left-normed n-weight commutators

$$[[...[g_1,g_2],...], g_n] \ (= [g_1,g_2,...,g_n]), \ g_i \in G,$$

where the commutator [g,h] is defined to be $g^{-1}h^{-1}gh$. Then $\gamma_n(G)$ is a fully invariant subgroup of G known as the *n-th lower central subgroup* of G. An important property of the dimension subgroup $D_n(G)$, $n \geq 1$, is that it contains the lower central subgroup $\gamma_n(G)$. To see this, first observe that in the group ring ZG,

$$[g,h] - 1 = g^{-1}h^{-1}gh - 1 = g^{-1}h^{-1}(gh - hg)$$

$$= g^{-1}h^{-1}((g-1)(h-1) - (h-1)(g-1)) \in \Delta^2(G),$$

so $\gamma_2(G) \leq D_2(G)$. Since $\gamma_n(G)$ is generated by all left-normed commutators $[g_1,g_2,...,g_n]$, $g_i \in G$, a simple induction on $n \geq 2$ yields the relation

$$[g_1,g_2,...,g_n] - 1 \in \Delta^n(G)$$

from which the general inequality $\gamma_n(G) \leq D_n(G)$ follows trivially. It is not difficult to see that $D_2(G) = \gamma_2(G)$. For instance, if $g = x_1^{e(1)}...x_k^{e(k)}$, with x_i independent, then modulo $\Delta^2(G)$, expansion of g-1 yields

$$g-1 \equiv e(1)(x_1-1) + ... + e(k)\,(x_k-1),$$

and if $g-1 \in \Delta^2(G)$ then we must have each $e(i) = 0$ implying $g = 1$. In these lectures I shall give a historical sketch of the developments of the so-called *dimension subgroup problem* which consists of the investigation of possible connections which exist between the subgroups $D_n(G)$ and $\gamma_n(G)$ for $n \geq 2$.

2. Dimension subgroups of free groups

If F is a free group then a well-known theorem of Magnus states that the equality $D_n(F) = \gamma_n(F)$ holds for all n. This result is known as the *fundamental theorem of free group rings*. It began as a conjecture by Magnus (1935) in the language of formal power series. Magnus (1937) gave a combinatorial proof of the conjecture. A minor gap in Magnus' proof was filled in by an identity supplied by Witt (1937) who, in turn, supplied another proof along the lines of Magnus (see,

Chandler and Magnus (1982)). Grün (1936) used an integral matrix representation of $F/\gamma_n(F)$ and gave another proof. Details of Grün's argument are given in a very informative article by Frank Röhl (1985) who has pointed out serious gaps in Grün's paper. Several accessible proofs of this important theorem are now available. We refer the reader to Gupta (1987). [Jennings (1941) proved that $D_{n,Q}(G)$, the n-th dimension subgroup of G over the rationals, coincides with $\sqrt{\gamma_n(G)}$ for all n and all G. It follows that if G is a group all of whose lower central factors are torsion-free then $D_n(G) = \gamma_n(G)$ for all n. Since free groups are known to have this property (which is not an easy fact to prove), the fundamental theorem also follows from this (see, Passi 1979).]

We recall Magnus embedding of the free group $F = \langle x_1, x_2, \ldots \rangle$ into the ring

$$\mathbf{P} = \mathbf{Z}[[a_1, a_2, \ldots]]$$

of formal power series over **Z** via the map $\beta : x_i \to 1 + a_i$, with

$$(1+a_i)^{-1} = 1 - a_i + a_i^2 - a_i^3 + \ldots .$$

The map β extends to a homomorphism of F into the group of units of **P** and it turns out to be a monomorphism.

[To see this, let

$$w = x_{i(1)}^{e(1)} \ldots x_{i(m)}^{e(m)}, \quad e(j) > 0, \quad x_{i(j)} \neq x_{i(j+1)},$$

be freely reduced in F. Then in the power series $\beta(w)$ the coefficient of $a_{i(1)} \ldots a_{i(m)}$ is easily seen to be $e(1) \ldots e(m)$ which is non-zero.]

Further, we extend β by linearity to $\beta : \mathbf{Z}F \to \mathbf{P}$. Following Magnus we define, for each $n \geq 1$,

$$\mathfrak{D}_n(F) = \{w \in F;\ \beta(w-1) \in \mathbf{J}^n\},$$

where **J** is the fundamental ideal of **P** (i.e. generated by $a_1, a_2, \ldots$).

Then it is clear that $\beta\Delta(F) \leq \mathbf{J}$ and it follows that $\beta\Delta^n(F) \leq \mathbf{J}^n$ implying, in turn, that

$$D_n(F) \leq \mathfrak{D}_n(F)$$

for all n.

Given an arbitrary element $u = u(x_1,...,x_m) \in \mathbf{Z}F$, we can express $u - \varepsilon(u)$ as

$$u - \varepsilon(u) = \sum_i (x_i-1)\partial_i(u), \qquad (*)$$

where $\varepsilon : \mathbf{Z}F \to \mathbf{Z}$ is the augmentation map.

The right coefficients $\partial_i(u)$ are the well-known right partial derivatives of u and (*) is known as Fox's fundamental formula.

[Alternately, $\partial_i(u)$ may be defined as the linear extensions of $\partial_i : F \to F$ defined by

$$\partial_i(x_i^{e_i}) = e_i x_i^{(e_i-1)/2} = \{1 \text{ if } e_i = 1; -x_i^{-1} \text{ if } e_i = -1\};$$

$$\partial_j(x_i^{e_i}) = 0, j \neq i;$$

and for $w = x_1^{a_1} \dots x_m^{a_m}$ $(a_j = \pm 1)$,

$$\partial_i(w) = a_i\, x_i^{(a_i-1)/2}\, x_{i+1}^{a_{i+1}} \dots x_m^{a_m}$$

where in w the x_j are not necessarily distinct.]

An iteration of Fox's fundamental formula shows that for $u = u(x_1,...,x_m) \in \mathbf{Z}F$,

$$\begin{aligned} u &= \varepsilon(u) + \sum_i (x_i - 1)\partial_i(u) \\ &= \varepsilon(u) + \sum_i (x_i - 1)\varepsilon\partial_i(u) + \sum_{i,j} (x_i - 1)(x_j - 1)\partial_i\partial_j(u) \\ &= \varepsilon(u) + \sum_i (x_i - 1)\varepsilon\partial_i(u) + \sum_{i,j} (x_i - 1)(x_j - 1)\varepsilon\partial_i\partial_j(u) \\ &\quad + \sum_{i,j,k} (x_i - 1)(x_j - 1)(x_k - 1)\partial_i\partial_j\partial_k(u) \\ &= \dots . \end{aligned} \qquad (**)$$

In particular, $u \in \Delta^n(F)$ if and only if $\varepsilon\partial_{i(1)} \dots \partial_{i(k)}(u) = 0$ for all sequences $(i(1),...,i(k))$ and all $1 \leq k \leq n-1$. Applying β to (**) shows that if $\beta(u) \in J^n$ then

$$\varepsilon\partial_{i(1)} \dots \partial_{i(k)}(u) = 0 \text{ for all } k \leq n-1$$

and all sequences $(i(1),...,i(k))$ and conversely.

Thus, if $w \in \mathfrak{D}_n(F)$ then

$$\beta(w-1) \in \mathbf{J}^n$$

and it follows that

$$\varepsilon\partial_{i(1)} \ldots \partial_{i(k)}(w-1) = 0 \text{ for all } 1 \leq k \leq n-1.$$

This implies that $w-1 \in \Delta^n(F)$ and consequently,

$$\mathfrak{D}_n(F) \leq D_n(F).$$

Thus, for free groups the two formulations of the dimension subgroup conjecture are the same (cf. Passi 1984). Translation by Fox (1953) in the language of group rings also allows an easy formulation of the general conjecture.

3. The dimension subgroup conjecture

The so-called *dimension subgroup conjecture* states that the equality $D_n(G) = \gamma_n(G)$ holds for all G and all n. While this conjecture seems to have been stimulated by Magnus (1935), for arbitrary groups it is not connected with his "dimension subgroup conjecture" as Magnus was not concerned with the ideals of group rings (for free groups, however, the two conjectures are the same as has been pointed out earlier). The first reference to the group ring version of the conjecture appears to be in Grün (1940) who actually uses the conjecture as *Magnus' theorem.* Since then the conjecture has eluded many researchers and has had a disturbing history (e.g. Cohn (1952), Losey (1960) [see Lyndon, Math. Reviews **26**, #6260], Fox (1953) and others).

[Cohn's result does however imply that the conjecture holds for groups of prime exponent (see Passi 1968 for correctional comments on Cohn's and Losey's papers). Fox did not actually attempt to resolve the conjecture. He just made a passing remark as to its validity.]

If the conjecture is indeed false then it is already false for some finite p-group. To see this, let G be a group such that for some $n \geq 1$, $\gamma_n(G) = 1$ whereas $D_n(G) \neq 1$. Then there is an element g in G such that $g \neq 1$, and g-1 is a **Z**-linear sum of elements of the form

$$(h_1-1)\ldots(h_n-1)h_{n+1},\ h_i \in G.$$

Let H be the subgroup of G generated by g together with the elements h_i of G which occur in the support of g-1. Then H is a finitely generated group with the property that $\gamma_n(H) = 1$ whereas $D_n(H) \neq 1$. Being a finitely generated nilpotent group, H is residually a finite p-group (Gruenberg 1957). It follows that for some finite p-group H/K,

$$D_n(H/K) \neq \gamma_n(H/K).$$

This observation has been attributed to Graham Higman (see Passi 1968). The above proof is adapted from Passi (1968). Sandling (1972) notes that, in addition, the centre of H/K can be assumed to be cyclic.

The equality $D_3(G) = \gamma_3(G)$ is due to Higman and Rees idependently (see Passi 1968), a proof of this is included in Section 4. Other proofs have since been given by Passi (1968), Hoare (1969), Sandling (1972), Bachman & Grünenfelder (1972), Losey (1974) and Gupta (1982). Passi (1968) was among the first who brought the dimension subgroup problem to limelight again. Among other things, he proved that if G is an odd p-group then $D_4(G) = \gamma_4(G)$, so a counter-example for n = 4 had to be a 2-group. Passi (1968) also obtained some positive results for the 2-group case. Still on the positive side of the conjecture, Moran (1970) proved that if G is a p-group then $D_n(G) = \gamma_n(G)$ for $n \leq p$. [Earlier, Jennings (1941) had proved the p-group conjecture for groups of exponent p. Passi, Sucheta and Tahara (1986) point out how Moran's result implies Jenning's result: $D_{n,Q}(G) = \sqrt{\gamma_n(G)}$.] Sandling (1972) also proved that the conjecture holds if G is an extension of an abelian group by a cyclic group (another proof is given by Passi (1974)). A counter-example to the dimension subgroup conjecture for n = 4 was given by Rips (1972) who constructed a finite 2-group G of order 2^{38} with $D_4(G) \neq \gamma_4(G)$. This result soon became a mile-stone in the development of the dimension subgroup problem. Tahara (1977) gave a complete description of the quotient $D_4(G)/\gamma_4(G)$. In particular, if G is generated by two or three elements then $D_4(G) = \gamma_4(G)$. Recently, Gupta (1989) has constructed counter-examples to the conjecture for each $n \geq 4$. The examples are described in Section 5.

Sandling (1972) proposes some modifications to the conjecture leading towards the so-called Lie dimension subgroups, and obtains some additional results.

Define $\Lambda(G) = \Delta(G)$, and for $n \geq 2$,

$$\Lambda^n(G) = ZG(\Lambda^{n-1}(G),\Lambda(G)),$$

where for subsets A, B of ZG, (A,B) is the additive subgroup of ZG generated by all Lie elements

$$(a,b) = ab-ba,$$

$a \in A$, $b \in B$. For each $n \geq 1$, define the n-*th Lie dimension subgroup* $LD_n(G)$ of G by

$$LD_n(G) = G \cap (1 + \Lambda^n(G)) = \{g \in G; g-1 \in \Lambda^n(G)\}.$$

Since $\Lambda^n(G) \leq \Delta^n(G)$ and $[g_1,...,g_n] - 1 \in \Lambda^n(G)$ (by an easy induction on n) it is clear that

$$\gamma_n(G) \leq LD_n(G) \leq D_n(G).$$

A weaker form of the dimension subgroup problem is the following:

Lie dimension subgroup problem:

Identify the quotient $LD_n(G)/\gamma_n(G)$ for all G and all n.

In this connection Sandling (1972) proved the following results:

(a) $LD_n(G) = \gamma_n(G)$ for all G and all $n \leq 6$;

(b) $LD_n(G) = \gamma_n(G)$ for all n if G is a p-group with $\gamma_p(G') = \{1\}$;

(c) $LD_n(G) = \gamma_n(G)$ for all $n \leq 2p$ if G is a p-group;

(d) $LD_n(G) = \gamma_n(G)$ for all n if G is metabelian.

[See Remark 2 in Section 6 for an alternate proof of (d).]

Remark 1. Passi and Sehgal (1975) have identified Lie dimension subgroups over fields and conclude, in particular: $LD_n(G) = \gamma_n(G)$ for all n if G' has exponent p. Recently Hurley and Sehgal (1989) have announced that for $n \geq 9$, $LD_n(G) \neq \gamma_n(G)$ in general. Their construction is modelled on the construction given in Gupta (1989). It will be of interest to known what happens when n = 7 or 8.

Remark 2. For each $n \geq 1$, define

$$\Omega^n(G) = \mathbf{Z}G(G,G,...,G) \text{ (G repeats n times)}$$

to be the ideal of $\mathbf{Z}G$ generated by all n-weight left-normed Lie elements

$$(g_1, g_2, ..., g_n),\ g_i \in G,$$

where $(g,h) = gh-hg$. Then, in general, $\Omega^n(G) \neq \Lambda^n(G)$. The following result is of independent interest:

Theorem (Gupta-Levin 1982). *For any group* G,

$$\gamma_n(G) \leq G \cap (1 + \Omega^n(G))$$

for all n.

4. The dimension subgroup problem

In view of Rips' counter-example the dimension subgroup problem assumes the following intepretation:

Problem 1. Identify the structure of the quotient group $D_n(G)/\gamma_n(G)$ for $n \geq 4$.

When $n = 4$, Losey (1974) proved that $D_n(G)/\gamma_n(G)$ has exponent dividing 2. Other proofs of this result have been given by Tahara (1977), Sjogren (1979) and Passi (1979). Tahara solved Problem 1 completely for the case $n = 4$, pointing out all possible counter-examples to the conjecture. An alternate description of $D_4(G)/\gamma_4(G)$ has been given by Gupta (1984) (see Gupta (1987), Theorem IV.5.1). The next important contribution towards the dimension subgroup problem is due to Sjogren (1979). The main result is that the exponent of the quotient $D_n(G)/\gamma_n(G)$ is bounded by a certain complicatedly defined number c(n). The number c(n) is independent of the group G and the prime factorization of c(n) uses primes not exceeding n-2. The first few c(i) are:

$$c(1) = 1,\ c(2) = 1,\ c(3) = 1,\ c(4) = 2,\ c(5) = 48.$$

Sjogren's proof seems to have been inspired by Stallings (1975) and is homological in nature. It uses the terminology of spectral sequences and the methods of Chen, Fox & Lyndon (1958). A modified account of Sjogren's proof was first given by Hartley (1982) who, among other things, unfolded c(n) to be:

$$c(n) = b(1)^{\binom{n-2}{1}} \cdots b(n-2)^{\binom{n-2}{n-2}},$$

where $b(k) = \ell.c.m.\ \{1,\ldots,k\}$. Further modifications of the argument by Cliff & Hartley (1985) and Gupta (1985, unpublished notes) yield a very accessible proof of Sjogren's theorem. Full details can now be found in Gupta (1987). We give an outline of the strategy used.

It is instructive to change to the language of free groups and free group rings. Let G be a finitely generated group given by the presentation

$$G = < x_1, \ldots, x_m;\ r_1, r_2, \ldots >.$$

Then $G \cong F/R$ where $F = < x_1,\ldots,x_m >$ is free and $R = < r_1, r_2, \ldots >^F$ is the normal closure of the relators $r_1, r_2, \ldots$.

Let

$$\mathfrak{f} = ZF(F-1)$$

be the augmentation ideal of **ZF** and

$$\mathfrak{r} = ZF(R-1)$$

be the augmentation ideal of **ZF** with respect to the normal subgroup R of F. Then in the language of free group rings the *dimension subgroup problem* (Problem 1) assumes the form:

Problem 1*. Identify the normal subgroup $F \cap (1 + \mathfrak{r} + \mathfrak{f}^n)$.

Note that $R\gamma_n(F) \leq F \cap (1 + \mathfrak{r} + \mathfrak{f}^n)$ and that the dimension subgroup conjecture amounts to the equality

$$F \cap (1 + \mathfrak{r} + \mathfrak{f}^n) = R\gamma_n(F)$$

for all normal subgroups R of F and all n. We define ideals $\mathfrak{r}(k)$, $k \geq 1$, of the free group ring **ZF** as follows:

$$\mathfrak{r}(1) = \mathfrak{r} = ZF(R-1) \leq \mathfrak{f},$$

and more generally, for $k \geq 2$,

$$\mathfrak{r}(k) = \sum_{i+j=k-1} \mathfrak{f}^i \mathfrak{r} \mathfrak{f}^j.$$

Also define $R(1) = R$, and more generally, for $k \geq 2$,

$$R(k) = [R,F,\ldots,F] \text{ (F repeats k-1 times)},$$

the subgroup generated by all left-normed commutators of the form $[r,f_2,\ldots,f_k]$ where $r \in R$ and $f_i \in F$. Sjogren's key result is the following:

Lemma 4.1 (Sjogren). *Let* $w \in \gamma_n(F)$, $n \geq 2$, *be such that*

$$w-1 \in \mathfrak{r}(k) + \mathfrak{f}^{n+1},$$

for some $2 \leq k \leq n$. *Then*

$$w^{b(k)}-1 \equiv f_k-1 \text{ modulo } \mathfrak{r}(k+1) + \mathfrak{f}^{n+1},$$

where $f_k \in R(k)$ *and* $b(k) = \ell.c.m.\{1,\ldots,k\}$.

[See Gupta (1987), IV.1.5 (A).]

When $k = n$, Lemma 4.1 admits a much sharper conclusion.

Lemma 4.2 (Sjogren). *Let* $w \in \gamma_n(G)$, $n \geq 2$, *be such that*

$$w-1 \in \mathfrak{r}(n) + \mathfrak{f}^{n+1}.$$

Then $w-1 \equiv f-1$ modulo $\mathfrak{f}^{n+1}$, *where* $f \in R(n)$.

[See Gupta (1987), IV.1.5(B).]

To deduce Sjogren's theorem we also need the following lemma whose foundation lies in Sjogren (1979) and Hartley (1982).

Lemma 4.3 (Gupta 1985). *Let* $H = H_1 \geq H_2 \geq \ldots$ *and* $K = K_1 \geq K_2 \geq \ldots$ *be series of normal subgroups of a group* F *(not necessarily free) and let*

$$\{D_{k,\ell};\ 1 \leq k \leq \ell\}$$

be a family of normal subgroups of F *such that*

(a) $D_{k,k+1} = H_k K_{k+1}$;

(b) $H_k K_\ell \leq D_{k,\ell}$;

(c) $D_{k,\ell+1} \leq D_{k,\ell}$ *for all* $k < \ell$.

If for each $2 \le k+m \le n+1$, $k,m,n \ge 1$, *there exists a positive integer* $a(k)$ *(depending on* k *and* n*) such that*

$$(K_{k+m} \cap D_{k,k+m+1})^{a(k)} \le D_{k+1,k+m+1} H_k,$$

then

$$(D_{1,n+2})^{a(1,n+1)} \le H_1 K_{n+2},$$

where

$$a(1,n+1) = a(1)^{\binom{n}{1}} \cdots a(n)^{\binom{n}{n}}.$$

[See Gupta (1987), IV.1.6.]

We can now deduce the following important theorem of Sjogren.

Theorem 4.4 (Sjogren 1979). *For all groups* G, $D_n(G)/\gamma_n(G)$, $n \ge 3$, *has exponent dividing*

$$c(n) = b(1)^{\binom{n-2}{1}} \cdots b(n-2)^{\binom{n-2}{n-2}},$$

where $b(k) = \ell.c.m.\{1,\dots,k\}$.

Proof. Let $G \cong F/R$ be given and let $n \ge 3$ be fixed. Let $w \in F \cap (1 + \mathfrak{r} + \mathfrak{f}^n)$. Then the proof consists in showing that $w^{c(n)} \in R\gamma_n(F)$.

Set

$$H_k = R(k),\ k \ge 1; \quad K_\ell = \gamma_\ell(F),\ \ell \ge 1,$$

and

$$D_{k,\ell} = F \cap (1 + \mathfrak{r}(k) + \mathfrak{f}^\ell).$$

Then

$$D_{k,k+1} = H_k K_{k+1} \quad \text{(by Lemma 4.2)},$$

$$H_k K_\ell \le D_{k,\ell}$$

and

$$D_{k,\ell+1} \le D_{k,\ell} \text{ for all } k \le \ell.$$

In addition, by Lemma 4.1,

$$(K_{k+1} \cap D_{k,k+m+1})^{b(k)} \leq D_{k+1,k+m+1} H_k,$$

for all $2 \leq k+m \leq n$. Thus by Lemma 4.3 it follows that $D_{1,n}/H_1K_n$ has exponent dividing

$$c(n) = b(1)^{\binom{n-2}{1}} \dots b(n-2)^{\binom{n-2}{n-2}}.$$

Since $D_{1,n} = F \cap (1 + \mathfrak{r} + \mathfrak{f}^n)$, $H_1 = R$ and $K_n = \gamma_n(F)$, it follows that the quotient

$$F \cap (1 + \mathfrak{r} + \mathfrak{f}^n)/R\gamma_n(F)$$

has exponent dividing c(n).

An important consequence of Sjogren's theorem is:

Corollary 4.5. *If* G *is a* p-*group then* $D_n(G) = \gamma_n(G)$ *for* $n \leq p+1$.

[Several earlier known results follow as consequences of Sjogren's theorem: $D_3(G) = \gamma_3(G)$ (Higman-Rees); $D_4(G)/\gamma_4(G)$ has exponent dividing 2 (Losey); $D_4(G) = \gamma_4(G)$ if G is an odd p-group (Passi); $D_n(G) = \gamma_n(G)$ for $n \leq p$ (Moran); $D_n(G) = \gamma_n(G)$ for all n if the lower central factors of G are all torsion-free (Hall-Jennings).]

[The bound c(n) for the exponent of $D_n(G)/\gamma_n(G)$ is, however, very crude. For instance when n = 5, c(n) = 48 whereas Tahara (1981) has proved directly that $D_5(G)/\gamma_5(G)$ has exponent dividing 6.]

Let w be an arbitrary element of $F \cap (1 + \mathfrak{r} + \mathfrak{f}^n)$. Then $w - 1 \in \mathfrak{r} + \mathfrak{f}^n$, and we obtain, in turn, the congruences:

$$w - 1 \equiv r - 1 \bmod \mathfrak{f}\mathfrak{r} + \mathfrak{f}^n, \text{ for some } r \text{ in } R;$$

$$wr^{-1} - 1 \equiv 0 \text{ modulo } \mathfrak{f}\mathfrak{r} + \mathfrak{f}^n.$$

Thus $wr^{-1} \in F \cap (1 + \mathfrak{f}\mathfrak{r} + \mathfrak{f}^n)$. If we can prove that

$$F \cap (1 + \mathfrak{f}\mathfrak{r} + \mathfrak{f}^n) \leq R\gamma_nF$$

then it would follow that $wr^{-1} \in R\gamma_n(F)$ and, in turn, $w \in R\gamma_n(F)$. Thus we have proved the following reduction lemma:

Lemma 4.6. $F \cap (1 + \mathfrak{r} + \mathfrak{f}^n) = R\gamma_n(F)$ *if and only if*

$$F \cap (1 + \mathfrak{f}\mathfrak{r} + \mathfrak{f}^n) \leq R\gamma_n(F).$$

In view of the fundamental theorem of free group rings: $F \cap (1 + \mathfrak{f}^n) = R\gamma_n(F)$ and the classical theorem of Magnus (1939): $F \cap (1 + \mathfrak{f}\mathfrak{r}) = R'$, the commutator subgroup of R, Lemma 4.6 is most tantalizing. For, while the identification of

$$F \cap (1 + \mathfrak{f}\mathfrak{r} + \mathfrak{f}^n)$$

should amount to $R'\gamma_n(F)$, for the purpose of the dimension subgroup conjecture we are asking the following much weaker question:

Problem 2. When is $F \cap (1 + \mathfrak{f}\mathfrak{r} + \mathfrak{f}^n) \leq R\gamma_n(F)$?

As an application of the above reduction lemma we conclude this section with a proof of the dimension subgroup conjecture for n = 3.

Theorem 4.7. $D_3(G) = \gamma_3(G)$ *for all* G.

Proof. Let G be given by a pre-abelian presentation

$$G = \langle x_1,\ldots,x_m;\ r_1,r_2,\ldots \rangle,$$

where

$$r_1 = x_1^{e(1)}\xi_1,\ldots,\ r_m = x_m^{e(m)}\xi_m,\ r_{m+1} = \xi_{m+1},\ \ldots,\qquad \xi_i \in F',$$

$$e(m) \mid e(m-1) \mid \ldots \mid e(1) \geq 0.$$

[Every finitely presented group admits such a presentation. See, for instance, Magnus *et al.* (1966), page 140.]

Then $G \cong F/R$ where

$$F = \langle x_1,\ldots,x_m \rangle$$

is free and

$$R = \langle r_1,r_2,\ldots \rangle^F$$

is the normal closure of the relators r_i. Let

$$S = \langle x_1^{e(1)},\ldots,\ x_m^{e(m)} \rangle^{F'}$$

and define

$$\mathfrak{s} = ZF(S-1).$$

Then

$$F \cap (1 + \mathfrak{fr} + \mathfrak{f}^3) \leq F \cap (1 + \mathfrak{fs} + \mathfrak{f}^3).$$

On the other hand

$$F \cap (1 + \mathfrak{fr} + \mathfrak{f}^3) \leq F \cap (1 + \mathfrak{f}^2) = F'.$$

Thus if $w \in F \cap (1 + \mathfrak{fr} + \mathfrak{f}^3)$ then

$$w = \prod_{1 \leq i < j \leq m} [x_i, x_j]^{a(i,j)}\, w', \tag{*}$$

$a(i,j) \in \mathbf{Z}$, $w' \in \gamma_3(F)$, and $w\text{-}1 \equiv 0 \bmod \mathfrak{fs} + \mathfrak{f}^3$.

Expansion of w-1 shows that

$$w\text{-}1 \equiv \sum_{1 \leq i < j \leq m} a(i,j) \{(x_i\text{-}1)(x_j\text{-}1) - (x_j\text{-}1)(x_i\text{-}1)\} \bmod \mathfrak{fs} + \mathfrak{f}^3.$$

Since $w\text{-}1 \equiv 0 \bmod \mathfrak{fs} + \mathfrak{f}^3$ and the ideals $\mathfrak{s}, \mathfrak{f}$ are invariant under the endomorphisms sending x_i to 1 and x_j to x_j for $j \neq i$, it follows that we must have

$$a(i,j)\{(x_i\text{-}1)(x_j\text{-}1) - (x_j\text{-}1)(x_i\text{-}1)\} \equiv 0 \bmod \mathfrak{fs} + \mathfrak{f}^3,$$

for each $i < j$. Further, since $\mathfrak{f}$ is a free right ZF-module (see, for instance, Gupta (1987), I.1.11), it follows that

$$a(i,j)(x_i\text{-}1)(x_j\text{-}1) \equiv 0 \bmod \mathfrak{fs} + \mathfrak{f}^3$$

for each $i < j$, and this in turn yields

$$a(i,j)(x_i\text{-}1) \equiv 0 \bmod \mathfrak{s} + \mathfrak{f}^2,$$

which is possible if and only if $a(i,j) = e(i)b(i,j)$ for some positive integer $b(i,j)$.

Substituting in the expression for w given by (*) yields

$$w = \prod_{1 \leq i < j \leq m} [x_i, x_j]^{a(i,j)} w' = \prod_{1 \leq i < j \leq m} [x_i^{e(i)}, x_j]^{b(i,j)}\, w'',$$

$w'' \in \gamma_3(F)$. Since $x_i^{e(i)} \xi_i$ is in R, it follows that w lies in $R\gamma_3(F)$ as was to be proved.

The last sentence in the proof of Lemma 4.7 shows that in fact w lies in $[R,F]\gamma_3(F)$. Thus we have proved:

Theorem 4.8. $F \cap (1 + \mathfrak{f}\mathfrak{r} + \mathfrak{f}^3) \leq [R,F]\gamma_3(F)$.

5. Counter-examples to the conjecture

As has been remarked earlier, for G = F/R the dimension subgroup conjecture amounts to the equality: $F \cap (1 + \mathfrak{r} + \mathfrak{f}^n) = R\gamma_n(F)$. In what follows we shall exhibit, for each $n \geq 4$, a normal subgroup R of F such that

$$F \cap (1 + \mathfrak{r} + \mathfrak{f}^n) \neq R\gamma_n(F).$$

For complete details the reader is referred to Gupta (1989).

Let $F = \langle x,a,b,c \rangle$ be a free group and let $n \geq 4$ be fixed. Define the normal subgroup

$$N_1 =$$

$$\langle F'', [x,a,b],[x,b,a],[x,a,c],[x,c,a],[x,b,c],[x,c,b],[x,a,x],[x,b,x],[x,c,x] \rangle^F,$$

where F" is the second commutator subgroup of F and [x,a,b] = [[x,a],b] etc. are left-normed commutators.

Observation 1. F/N_1 is a metabelian group in which $\gamma_3(F)/N_1$ is a free abelian subgroup freely generated by the cosets

$$[x,(k)a]N_1,\ [x,(k)b]N_1,\ [x,(k)c]N_1,\ k \geq 2,$$

where for any $y \in F$, $[x,(1)y] = [x,y]$ and inductively,

$$[x,(i)y] = [[x,(i-1)y],y],\ i \geq 2.$$

Define the normal subgroup

$$N_2 = \langle N_1,\ [x,(n-1)a],[x,(n-1)b],[x,(n-1)c] \rangle^F.$$

Observation 2. F/N_2 is nilpotent of class precisely n-1.

Define the normal subgroup

$$N_3 = \langle N_2,[x,a^{2^{n+2}}],[x,b^{2^n}],[x,c^{2^{n-2}}] \rangle^F.$$

Observation 3. Each lower central factor $\gamma_k(F)N_3/\gamma_{k+1}(F)N_3$, $k \geq 2$, is a finite 2-group of order 2^{3n}; moreover the precise orders of

$$[x,(k)a]N_3,\ [x,(k)b]N_3,\ [x,(k)c]N_3,\ k \geq 1,$$

are given by the congruences

$$[x,(k)a]^{2^{2n-k}} \equiv 1,\ [x,(k)b]^{2^{2n-k-2}} \equiv 1,\ [x,(k)c]^{2^{2n-k-4}} \equiv 1 \pmod{N_3}.$$

This can be proved by reverse induction on $k = n-2,\ldots,2$, using the congruences

$$[x,a^{2^{n+2}},(k-1)a] \equiv 1,\ [x,b^{2^n},(k-1)b] \equiv 1,\ [x,c^{2^{n-2}},(k-1)c] \equiv 1 \pmod{N_3}.$$

For instance, the precise order 2^{n+2} of $[x,(n-2)a]N_3$ comes from the congruence

$$[x,a^{2^{n+2}},(n-3)a] \equiv 1 \pmod{N_3}.$$

Define the subgroup

$$N_4 = < N_3,\ [x,(n-2)c][x,(n-2)b]^{-4},\ [x,(n-2)b][x,(n-2)a]^{-4} >.$$

Observation 4. $[N_4,F] \leq N_3$. In particular, N_4 is normal in F and $\gamma_{n-2}(F)/N_4$ is cyclic of order precisely 2^{n+2} generated by the coset zN_4, where $z = [x,(n-2)a]$.

Define the subgroup

$$N_5 = < N_4,\ [a,b]^{2^n} z^{-4},\ [b,c]^{2^{n-2}} z^{-4},\ [c,a]^{2^{n-2}} z^2 >.$$

Observation 5. $[N_5,F] \leq N_4$, in particular N_5 is normal in F. In addition, the orders of $[a,b]N_5$, $[b,c]N_5$, $[c,a]N_5$ are respectively 2^{2n}, 2^{2n-2}, 2^{2n-1}.

[This follows using the fact that $[a,b]N_4$, $[b,c]N_4$, $[c,a]N_4$, zN_4 are all in the centre of F/N_4 and the order of zN_4 is 2^{n+2}.]

Finally, define the subgroup

$$R = < N_5,\ u,v,w >,$$

where

$$u = a^{2^{n+2}}[x,(n-3)b]^{-4}[x,(n-3)c]^{-2},$$

$$v = b^{2^n}[x,(n-3)a]^4[x,(n-3)c]^{-1},$$

$$w = c^{2^{n-2}}[x,(n-3)a]^2[x,(n-3)b].$$

Observation 6. $[R,F] \leq N_5$ and, in particular, R is normal in F.

[This follows from the fact that each of [u,x], [u,a], [u,b], [u,c], [v,x], [v,a], [v,b], [v,c], [w,x], [w,a], [w,b], [w,c] lies in N_5.]

This completes the construction of R and all that remains is to prove that for this choice of R, $F \cap (1 + \mathfrak{r} + \mathfrak{f}^n) \neq R\gamma_n(F) = R$.

We shall need the following lemma of independent interest.

Lemma 7 (cf. Passi 1968). *Let* $m \geq 1$ *be fixed and let*

$$\mathfrak{f} = (x-1)\mathbf{Z}\langle x\rangle$$

and

$$\mathfrak{x} = (x^{2^m}-1)\mathbf{Z}\langle x\rangle$$

be ideals of the infinite cyclic group ring $\mathbf{Z}\langle x\rangle$. *Then*

$$2^{m+k}\mathfrak{f} \leq \mathfrak{f}^{k+2} + \mathfrak{x}$$

for all $k \geq 0$.

Proof. It clearly suffices to prove by induction on $k \geq 0$ that

$$2^{m+k}\mathfrak{f} \leq 2^{m-1}\mathfrak{f}^{k+2} + \ldots + 2\mathfrak{f}^{m+k} + \mathfrak{f}^{m+k+1} + \mathfrak{x}\,. \tag{1}$$

A straightforward expansion of $(x^{2^m}-1)$ yields

$$(x^{2^m}-1) = 2^m(x-1) + \sum_{i=2}^{2^m} \binom{2^m}{i}(x-1)^i$$

$$\equiv 2^m(x-1) \bmod 2^{m-1}\mathfrak{f}^2 + \ldots + 2\mathfrak{f}^m + \mathfrak{f}^{m+1}$$

and it follows that

$$2^m\mathfrak{f} \leq 2^{m-1}\mathfrak{f}^2 + \ldots + 2\mathfrak{f}^m + \mathfrak{f}^{m+1} + \mathfrak{x}, \tag{2}$$

proving the assertion for k=0. For the inductive step, we assume the result for some $k \geq 0$. Then by the induction hypothesis we have

$$2^{m+k}\mathfrak{f} \leq 2^{m-1}\mathfrak{f}^{k+2} + \ldots + 2\mathfrak{f}^{m+k} + \mathfrak{f}^{m+k+1} + \mathfrak{x},$$

which upon multiplication with 2 gives

$$2^{m+k+1}\mathfrak{f} \leq 2^{m}\mathfrak{f}^{k+2} + \ldots + 2^2\mathfrak{f}^{m+k} + 2\mathfrak{f}^{m+k+1} + \mathfrak{x}. \tag{3}$$

Multiplying both sides of (2) by $\mathfrak{f}^{k+1}$ gives

$$2^m\mathfrak{f}^{k+2} \leq 2^{m-1}\mathfrak{f}^{k+3} + \ldots + 2\mathfrak{f}^{m+k+1} + \mathfrak{f}^{m+k+2} + \mathfrak{x},$$

which upon substituting in (3) yields the desired relation,

$$2^{m+k+1}\mathfrak{f} \le 2^{m-1}\mathfrak{f}^{k+3} + \dots + 2\mathfrak{f}^{m+k+1} + \mathfrak{f}^{m+k+2} + \mathfrak{x}.$$

Observation 8. (i) $(a^{2^{n+2}}-1)\mathfrak{f}^2 + \mathfrak{f}(a^{2^{n+2}}-1)\mathfrak{f} + \mathfrak{f}^2(a^{2^{n+2}}-1) \le \mathfrak{r} + \mathfrak{f}^n$;

(ii) $(b^{2^n}-1)\mathfrak{f}^2 + \mathfrak{f}(b^{2^{2n}}-1)\mathfrak{f} + \mathfrak{f}^2(b^{2^{2n}}-1) \le \mathfrak{r} + \mathfrak{f}^n$;

(iii) $(c^{2^{n-2}}-1)\mathfrak{f}^2 + \mathfrak{f}(c^{2^{n-2}}-1)\mathfrak{f} + \mathfrak{f}^2(c^{2^{n-2}}-1) \le \mathfrak{r} + \mathfrak{f}^n$.

Details. Since $a^{2^{n+2}} \in R\gamma_{n-2}(F)$, it follows that $(a^{2^{n-2}}-1) \in \mathfrak{r} + \mathfrak{f}^{n-2}$, and the proof of (i) follows using Lemma 7. The details of (ii) and (iii) are similar.

Using Lemma 7 we can now deduce our key observation.

Observation 9. (i) $2^{2n-2}(a-1)(c-1) \equiv (a^{2^{2n-2}}-1)(c-1)$ modulo $\mathfrak{r} + \mathfrak{f}^n$;

(ii) $2^{2n-2}(a-1)(c-1) \equiv (a-1)(c^{2^{2n-2}}-1)$ modulo $\mathfrak{r} + \mathfrak{f}^n$;

(iii) $2^{2n-3}(b-1)(c-1) \equiv (b^{2^{2n-3}}-1)(c-1)$ modulo $\mathfrak{r} + \mathfrak{f}^n$;

(iv) $2^{2n-3}(b-1)(c-1) \equiv (b-1)(c^{2^{2n-3}}-1)$ modulo $\mathfrak{r} + \mathfrak{f}^n$;

(v) $2^{2n-1}(a-1)(b-1) \equiv (a^{2^{2n-1}}-1)(b-1)$ modulo $\mathfrak{r} + \mathfrak{f}^n$;

(vi) $2^{2n-1}(a-1)(b-1) \equiv (a-1)(b^{2^{2n-1}}-1)$ modulo $\mathfrak{r} + \mathfrak{f}^n$.

For our later use we shall also need the following observation which can be verified directly.

Observation 10. Each of the following three relations hold in F:

(i) $b^{2^{2n-1}}\, c^{2^{n-2}} \in \gamma_{n-2}(F)^{2^{n+2}}\, \gamma_{n-1}(F)R$;

(ii) $a^{2^{2n-1}}\, c^{-2^{2n-3}} \in \gamma_{n-2}(F)^{2^n}\, \gamma_{n-1}(F)R$;

(iii) $a^{2^{2n-2}}\, b^{2^{2n-3}} \in \gamma_{n-2}(F)^{2^{n-2}}\, \gamma_{n-1}(F)R$.

[For instance, working modulo $\gamma_{n-1}(F)R$, we obtain the congruences:

$$\begin{aligned} a^{2^{2n-1}}\, c^{-2^{2n-3}} &\equiv a^{2^{n+2}2^{n-3}}\, c^{-2^{n-2}2^{n-1}} \\ &\equiv ([x,(n-3)b]^4\, [x,(n-3)c]^2)^{2^{n-3}}\, ([x,(n-3)a]^2\, [x,(n-3)b])^{2^{n-1}} \\ &\equiv [x,(n-3)a]^{2^n}\, [x,(n-3)b]^{2^n} [x,(n-3)c]^{2^{n-2}} \\ &\equiv [x,(n-3)a]^{2^n}\, [x,(n-3)b]^{2^n}\, [x,(n-3)b]^{2^n} \in \gamma_{n-2}(F)^{2^n}\, \gamma_{n-1}(F).] \end{aligned}$$

The main result is our final observation.

Observation 11. (i) $[a,b]^{2^{2n-1}} \notin R\gamma_n(F) = R$;

(ii) $[a,b]^{2^{2n-1}} \in F \cap (1 + \mathfrak{r} + \mathfrak{f}^n)$.

Details. For the proof of (i) we note from Observation 5 that the order of [a,b]R is precisely 2^{2n}. For the proof of (ii) we first note by Observation 5 that the element $[a,c]^{2^{2n-2}}[b,c]^{2^{2n-3}} \equiv 1 \pmod R$. Thus it suffices to prove that

$$w = [a,b]^{2^{2n-1}}[a,c]^{2^{2n-2}}[b,c]^{2^{2n-3}} \in F \cap (1 + \mathfrak{r} + \mathfrak{f}^n).$$

Indeed, we use Observation 9 to obtain, in turn, the following congruences modulo $\mathfrak{r} + \mathfrak{f}$

$$\begin{aligned} w-1 &= ([[a,b]^{2^{2n-1}}[a,c]^{2^{2n-2}}[b,c]^{2^{2n-3}}-1) \\ &\equiv 2^{2n-1}\{(a-1)(b-1) - (b-1)(a-1)\} + 2^{2n-2}\{(a-1)(c-1) - (c-1)(a-1)\} \\ &\quad + 2^{2n-3}\{(b-1)(c-1) - (c-1)(b-1)\}; \\ &\equiv (a-1)(b^{2^{2n-1}}-1) - (b-1)(a^{2^{2n-1}}-1) + (a-1)(c^{2^{2n-2}}-1) - (c-1)(a^{2^{2n-2}}-1) \\ &\quad + (b-1)(c^{2^{2n-3}}-1) - (c-1)(b^{2^{2n-3}}-1); \\ &\equiv (a-1)(b^{2^{2n-1}}c^{2^{2n-2}}-1) - (b-1)(a^{2^{2n-1}}c^{-2^{2n-3}}-1) - (c-1)(a^{2^{2n-2}}b^{2^{2n-3}}-1). \end{aligned}$$

Thus by Observation 10, there exist $d_a, d_b, d_c \in R\gamma_{n-2}(F)$ such that modulo $\mathfrak{r}+\mathfrak{f}$,

$$\begin{aligned} w-1 &\equiv (a-1)(d_a^{2^{n+2}}-1) - (b-1)(d_b^{2^n}-1) - (c-1)(d_c^{2^{n-2}}-1) \\ &\equiv (a^{2^{n+2}}-1)(d_a-1) - (b^{2^n}-1)(d_b-1) - (c^{2^{n-2}}-1)(d_c-1) \quad (\text{since } n \geq 4) \\ &\equiv 0, \end{aligned}$$

as was to be shown.

Remark. It is easy to compute and verify that the order of F/R is 2^{3n^2-3} for each $n \geq 4$. Thus for n = 4, our group has order 2^{45} as compared with 2^{38}, the order of Rips' counter-example. It should be pointed out, however, that Rips' group is a large quotient of our group when n = 4.

6. Dimension subgroups of metabelian groups

Since the counter-examples to the dimension subgroup conjecture are already metabelian, it is significant to restrict to the study of the dimension subgroup problem for finitely generated metabelian groups. Let G be a finitely generated non-cyclic metabelian group given by a pre-abelian free presentation

$$1 \to R \to F \to G \to 1$$

such that

$$F = \langle x_1,\ldots,x_m;\ \varnothing \rangle,\ m \geq 2,$$

is free and R is the normal closure

$$R = \langle x_1^{e(1)}\zeta_1,\ldots, x_m^{e(m)}\zeta_m, \zeta_{m+1},\ldots, \zeta_s, F'' \rangle^F,$$

where $\zeta_i \in F'$ and $e(m) \mid e(m-1) \mid \ldots \mid e(1) \geq 0$. Set

$$S = \langle x_1^{e(1)},\ldots, x_m^{e(m)} \rangle F'.$$

Then F/S is a finitely generated abelian group. Further, we note that

$$S' = [S,S] \leq R.$$

Define

$$\mathbf{s} = \mathbb{Z}F(S-1).$$

Then $\mathbf{r} \leq \mathbf{s}$ and

$$F \cap (1 + \mathfrak{f}\mathbf{r} + \mathfrak{f}^n) \leq F \cap (1 + \mathfrak{f}\mathbf{s} + \mathfrak{f}^n).$$

From the reduction lemma we deduce:

Lemma 6.1. *Let* F/R *be a finitely generated metabelian group given by a pre-abelian presentation of the form above and let* S *be as defined above. If*

$$F \cap (1 + \mathfrak{f}\mathbf{s} + \mathfrak{f}^n) = S'\gamma_n(F)$$

then

$$F \cap (1 + \mathfrak{f}\mathbf{r} + \mathfrak{f}^n) \leq R\gamma_n(F).$$

Lemma 6.1 has a very special appeal to it. For, F/S is finitely generated abelian and in view of the identifications

$$F \cap (1 + \mathfrak{f}\mathbf{s}) = S' \text{ and } F \cap (1 + \mathfrak{f}^n) = \gamma_n(F),$$

it appears more plausible that the corresponding conjecture:

$$F \cap (1 + \mathfrak{f}\mathfrak{s} + \mathfrak{f}^n) = S'\gamma_n(F)$$

should be true. In view of the counter-examples, Lemma 6.1 implies that, in general,

$$F \cap (1 + \mathfrak{f}\mathfrak{s} + \mathfrak{f}^n) \neq S'\gamma_n(F).$$

It is natural to raise the following two problems:

Problem 3. For F/S abelian, when is $F \cap (1 + \mathfrak{f}\mathfrak{s} + \mathfrak{f}^n) = S'\gamma_n(F)$?

Problem 4. For F/S abelian, identify the subgroup $F \cap (1 + \mathfrak{f}\mathfrak{s} + \mathfrak{f}^n)$.

Let S be as before and let $w \in F \cap (1 + \mathfrak{f}\mathfrak{s} + \mathfrak{f}^n)$. Then $w \in F'$ and we may write w as

$$w \equiv \prod_{1 \le i < j \le m} [x_i, x_j]^{p(i,j)} \bmod F'' \qquad (*)$$

where $p(i,j) \in \mathbf{Z}<x_i, \ldots, x_m>$.

[This can be proved using the Jacobi identity: $[a,b,c][c,a,b][b,c,a] = 1$ which holds in metabelian groups, or see Bachmuth (1965).]

Expansion of w-1 modulo $\mathfrak{f}\mathfrak{s} + \mathfrak{f}^n$ yields the congruence

$$w-1 \equiv \prod_{1 \le i < j \le m} \{(x_i-1)(x_j-1) - (x_j-1)(x_i-1)\}p(i,j) \bmod \mathfrak{f}\mathfrak{s} + \mathfrak{f}^n.$$

Since $w-1 \equiv 0 \bmod \mathfrak{f}\mathfrak{s} + \mathfrak{f}^n$ and the ideals $\mathfrak{f}, \mathfrak{s}$ are invariant under the endomorphisms mapping x_i to 1 and x_j to x_j for $j \neq i$, it follows that for each $i \ge 1$,

$$\sum_{i<j \le m} \{(x_i-1)(x_j-1) - (x_j-1)(x_i-1)\}p(i,j) \equiv 0 \bmod \mathfrak{f}\,\mathfrak{s} + \mathfrak{f}^n.$$

Further, since $\mathfrak{f}$ is a free right ZF-module with basis $\{(x_i-1)\}$, we must have

$$(x_i-1)p(i,j) \equiv 0 \bmod \mathfrak{s} + \mathfrak{f}^{n-1},$$

for each $1 \le i < j \le m$.

Since $\mathfrak{s} = \mathfrak{a} + (x_1^{e(1)}-1)ZF + \ldots + (x_m^{e(m)}-1)ZF$, it follows that

$$p(i,j) \equiv 0 \bmod t_i(x_i)ZF + \mathfrak{s} + \mathfrak{f}^{n-2},$$

where

$$t_i(x_i) = 1 + x_i + \ldots + x_i^{e(i)-1} \quad (t_i(x_i) = 0 \text{ if } e(i) = 0).$$

Substituting the value of p(i,j) in (*) yields

$$w \equiv \prod_{1\le i<j\le m} [x_i,x_j]^{t_i(x_i)q(i,j)} \bmod S'\gamma_n(F),\ q(i,j) \in \mathbf{Z}F,$$

or equivalently,

$$w \equiv \prod_{1\le i<j\le m} [x_i^{e(i)},x_j]^{q(i,j)} \bmod S'\gamma_n(F). \qquad (**)$$

An immediate consequence of (**) is:

Theorem 6.2 (Gupta 1982). $F \cap (1 + \mathfrak{f}\,\mathfrak{s} + \mathfrak{f}^n) \le [S^*,F]S'\gamma_n(F)$, *where* $S^* = \langle x_1^{e(1)},\dots,x_m^{e(m)} \rangle^F$.

This simple observation is the starting point for the development of the dimension subgroup problem for metabelian groups. For instance, in particular, if $S^* = \{1\}$ then

$$F \cap (1 + \mathfrak{f}\,\mathfrak{s} + \mathfrak{f}^n) = S'\gamma_n(F)$$

and we have:

Theorem 6.3 (Gupta 1982). $F \cap (1 + \mathfrak{f}\mathfrak{a} + \mathfrak{f}^n) = F''\gamma_n(F)$.

Consequently:

If G *is a free metabelian group then* $D_n(G) = \gamma_n(G)$ *for all* n.

[Since the lower central factors of a free metabelian group are torsion free, this result also follows from the Hall-Jennings theorem mentioned earlier.]

Among other consequences of Theorem 6.2 are:

If G *is a metabelian group then the exponent of* $D_n(G)/\gamma_n(G)$ *divides the exponent of* G/G'.

If G *is a metabelian group then* $[D_n(G),G] = \gamma_{n+1}(G)$.

[See Section IV.3 of Gupta (1987) for proofs.]

Theorem 6.2 further reduces Problem 3 to the following problem:

Problem 3*. For F/S abelian, when is

$$[S^*,F] \cap (1 + \mathfrak{f}\,\mathfrak{s} + \mathfrak{f}^n) = S'\gamma_n(F)?$$

This problem has been partly resolved as follows:

Theorem 6.4 (Gupta-Hales-Passi 1983). *Let* F/S *be a finitely generated abelian group. Then there exists* $n_0 = n_0(F/S)$ *such that*

$$[S^*,F] \cap (1 + \mathfrak{f}\,\mathfrak{s} + \mathfrak{f}^n) = S'\gamma_n(F) \textit{ for all } n \geq n_0.$$

As a corollary we obtain:

If G *is a finitely generated metabelian group then there exists* n_0 *(depending only on the invariants of* G/G'*) such that* $D_n(G) = \gamma_n(G)$ *for all* $n \geq n_0$.

Remark 1. Problem 3* has been completely resolved for certain abelian p-groups F/S. For instance, Hales (1985) has shown that if F/S is an abelian p-group of exponent p^e then

$$F \cap (1 + \mathfrak{f}\,\mathfrak{s} + \mathfrak{f}^n) = S'\gamma_n(F) \text{ for all } n \geq p^e+p^{e-1},$$

and if $n < p^e+p^{e-1}$ then

$$F \cap (1 + \mathfrak{f}\,\mathfrak{s} + \mathfrak{f}^n) \neq S'\gamma_n(F).$$

Remark 2. In the language of free group rings it is easy to give an alternate proof of Sandling's theorem about the Lie dimension subgroups of metabelian groups, namely $LD_n(G) = \gamma_n(G)$ for all n if G is metabelian. The theorem takes the form:

$$F \cap (1 + \mathfrak{f}\mathfrak{r} + \mathbf{Z}F(\mathfrak{f}, \mathfrak{f}, \ldots, \mathfrak{f})) \leq R\gamma_n(F),$$

where $\mathfrak{f}$ repeats n times in the Lie ideal.

To see this we observe first that

$$\mathbf{Z}F(\mathfrak{f}, \mathfrak{f}, \ldots, \mathfrak{f}) \leq \mathbf{Z}(\gamma_n(F) - 1) + \mathfrak{f}\mathfrak{a}.$$

So,

$$\begin{aligned} &F \cap (1 + \mathfrak{f}\mathfrak{r} + \mathbf{Z}F(\mathfrak{f}, \mathfrak{f}, \ldots, \mathfrak{f})) \\ &\qquad \leq F \cap (1 + \mathfrak{f}\,\mathfrak{s} + \mathbf{Z}F(\mathfrak{f}, \mathfrak{f}, \ldots, \mathfrak{f})) \\ &\qquad \leq F \cap (1 + \mathfrak{f}\,\mathfrak{s} + \mathbf{Z}(\gamma_n(F)-1)). \end{aligned}$$

Thus $w-1 \in \mathfrak{f}\,\mathfrak{s} + Z(\gamma_n(F)-1)$ implies that $wu-1 \in \mathfrak{f}\,\mathfrak{s}$ for some u in $\gamma_n(F)$ and this in turn yields using Magnus (1939) that $wu \in S'$. Thus $w \in S'\gamma_n(F)$ which is contained in $R\gamma_n(F)$ as required.

Remark 3. If G is a finitely generated metabelian group then a direct proof of an improved version of Sjogren's theorem is possible. For instance, it can be proved directly that $D_n(G)/\gamma_n(G)$ has exponent dividing 2.b(1)b(2)...b(n-2), where b(k) is the least common multiple of {1,...,k}. [See, Gupta (1987), IV.4.6.]

For metabelian p-groups Gupta-Tahara (1985) proved the following result:

If G *is a metabelian* p-*group,* p *odd, then* $D_n(G) = \gamma_n(G)$ *for* $n \le p+2$.

[This is an improvement over Sjogren's result: $D_n(G) = \gamma_n(G)$ for $n \le p+1$ for arbitrary p-groups.]

Developing further the theory of group rings of metabelian groups with respect to Sjogren's theory, Gupta (1988) proved the following theorem which significantly improves the Gupta-Tahara result:

If G *is a metabelian* p-*group then* $D_n(G) = \gamma_n(G)$ *for* $n \le 2p-1$.

7. Solution of the problem for metabelian p-groups

The developments of the dimension subgroup problem reported earlier have led Gupta (1989) to completely resolve the problem for metabelian p-groups when p is an odd prime. The main conclusion is:

If G *is a metabelian* p-*group,* p *odd, then* $D_n(G) = \gamma_n(G)$ *for all n.*

We give an outline of the proof of this theorem. As before we define the ideals

$$\mathfrak{r}(k) = \sum_{i+j+l=k} \mathfrak{f}^i\,\mathfrak{r}\,\mathfrak{f}^j$$

and the normal subgroups $R(k) = [R(k-1),F]$, $k \ge 1$.

Let $\mathfrak{a} = ZF(F'-1)$, $F' = \gamma_2(F)$. We readily observe that

$$R(k) \le F \cap (1 + \mathfrak{r}(k));\ F'' \le F \cap (1 + \mathfrak{a}^2) \text{ and } \gamma_q(F) \le F \cap (1 + \mathfrak{f}^q).$$

For each $1 \le k < q$, define the higher dimension subgroups

$$D(k,q) = F \cap (1 + \mathfrak{r}(k) + \mathfrak{a}^2 + \mathfrak{f}^q).$$

[Note that $D(k,q) \ge R(k)F''\gamma_q(F)$.]

The dimension subgroup problem for G = F/RF" translates to the study of the quotient groups

$$F \cap (1 + ZF(RF''-1) + \mathfrak{f}^q)/RF''\gamma_q(F),$$

which in turn are directly influenced by the quotients $D(1,q)/RF''\gamma_q(F)$. Thus, for example, if $D(1,q) = RF''\gamma_q(F)$ then $D_q(G) = \gamma_q(G)$.

We need to recall the following result:

Lemma 7.1. (Sjogren (1979), Gupta (1988)).

$$(\gamma_{k+q}(F) \cap D(k, k+q+1))^{b(k)} \leq D(k+1, k+q+1)\, R(k)F'',$$

$k,q \geq 1$, $b(k) = \ell.c.m.\ \{1,\ldots,k\}$.

Assume that we have proved the following theorem.

Theorem A. *For* $n > k \geq 3$, $D(k,n) \leq D(k+1,n)\, R(k)F''$.

Using Theorem A and Lemma 7.1, we can prove:

Theorem 7.2. *For* $n \geq 3$, $D(2,n)^{\alpha(n)} \leq R(2)F''\gamma_n(F)$, *where* $\alpha(n) = 2^{n-3}$.

Proof. When n = 3,

$$D(2,n) = F \cap (1 + \mathfrak{r}(2) + \mathfrak{a}^2 + \mathfrak{f}^3) = F \cap (1 + \mathfrak{r}(2) + \mathfrak{f}^3) = R(2)\gamma_3(F),$$

by Theorem 4.8. For the inductive step, assume the result for some $n \geq 3$ and let $w \in D(2,n+1)$. Then $w \in D(2,n)$ and by the induction hypothesis

$$w^{2^{n-3}} \in R(2)F''\gamma_n(F)$$

which implies that for some $u \in R(2)F''$, $w^{2^{n-3}}u \in \gamma_n(F)$. Also, by hypothesis, $w^{2^{n-3}}u \in D(2,n+1)$. Thus

$$w^{2^{n-3}}u \in \gamma_n(F) \cap D(2,n+1).$$

By Lemma 7.1 it follows that

$$(w^{2^{n-3}}u)^2 \in D(3,n+1)R(2)F'' \leq R(2)F''\gamma_{n+1}(F),$$

by Theorem A. Since $u \in R(2)F''$, it follows that $w^{2^{n-2}} \in R(2)F''\gamma_{n+1}(F)$ as required.

[I thank I.B.S. Passi for suggesting the above argument which is different from my original proof and gives a much sharper bound: 2^{n-3} as compared with the original bound $2^{\binom{n-2}{2}}$.]

As consequences, we deduce our main results.

Theorem 7.3. *If Theorem A holds and* G *is a finite metabelian group then* $D_n(G)/\gamma_n(G)$, $n \geq 4$, *has exponent dividing* $\alpha(n) = 2^{n-3}$.

Theorem 7.4. *If Theorem A holds and* G *is a finitely generated metabelian* p-*group*, p *odd, then* $D_n(G) = \gamma_n(G)$ *for all* $n \geq 1$.

Now, if we assume that, for $n > k \geq 3$,

$$F \cap (1 + \mathfrak{r}(k) + \mathfrak{f}^2\mathfrak{a} + \mathfrak{f}^n) = R(k)F''\gamma_n(F).$$

Then we would have

$$\begin{aligned} D(k,n) &= F \cap (1 + \mathfrak{r}(k) + \mathfrak{a}^2 + \mathfrak{f}^n) \\ &\leq F \cap (1 + \mathfrak{r}(k) + \mathfrak{f}^2\mathfrak{a} + \mathfrak{f}^n) \\ &= R(k)F''\gamma_n(F) \\ &= (F \cap (1 + \mathfrak{f}^n))R(k)F'' \\ &\leq (F \cap (1 + \mathfrak{r}(k+1) + \mathfrak{a}^2 + \mathfrak{f}^n))R(k)F'' \\ &= D(k+1, n)R(k)F''. \end{aligned}$$

Thus, the proof of Theorem A reduces to proving the following:

Theorem B. *For* $n > k \geq 3$, $F \cap (1 + \mathfrak{r}(k) + \mathfrak{f}^2\mathfrak{a} + \mathfrak{f}^n) = R(k)F''\gamma_n(F)$.

In preparation for the proof of Theorem B, we introduce in ZF the ideals $\mathfrak{R}(k)$ and the corresponding Z-modules $\mathfrak{R}(k)^*$, $k \geq 1$, defined by

$$\mathfrak{R}(k) = \mathbf{Z}F(R(k)-1) \text{ and } \mathfrak{R}(k)^* = \mathbf{Z}(R(k)-1).$$

Since $\mathfrak{R}(k) \leq \mathfrak{R}(k)^* + \mathfrak{r}(k+1)$, we have:

If $z \in \mathfrak{R}(k)$ *then* $z \equiv g-1 \bmod \mathfrak{r}(k+1)$ *for some* $g \in R(k)$.

We next list some elementary congruences (see, for instance, Gupta (1988)).

Lemma 7.5. *For* $k \geq 3$, *the following congruences hold in* ZF:

(i) $\quad (f_1-1)(f_2-1)\ldots(f_k-1) \equiv (f_1-1)(f_{2\sigma}-1)\ldots(f_{k\sigma}-1) \bmod \mathfrak{f}\mathfrak{a}$,

for all $f_i \in F$ *and all permutations* σ *of* $\{2,\ldots,k\}$;

(ii) $(r-1)(f_2-1)(f_3-1)...(f_k-1)$

$$\equiv (f_2-1)(r-1)(f_3-1)...(f_k-1) \mod (\mathfrak{f}\mathfrak{a} + \mathfrak{R}(k)^* + \mathfrak{r}(k+1)),$$

for all $f_i \in F$, $r \in R$;

(iii) $(f_1-1)(f_2-1)(f_3-1)...(f_{t-1}-1)(r-1)(f_{t+1}-1)...(f_k-1)$

$$\equiv (f_2-1)(f_1-1)(f_3-1)...(f_{t-1}-1)(r-1)(f_{t+1}-1)...(f_k-1)$$

mod $(\mathfrak{f}\mathfrak{a} + \mathfrak{R}(k)^* + \mathfrak{r}(k+1))$, *for all* $f_i \in F$, $r \in R$ *and* $3 \le t \le k$;

(iv) $(f-1)(x^e\zeta-1) \equiv (f-1)(x^e-1) \mod \mathfrak{f}\mathfrak{a}$,

for all $f,x \in F$ *and* $\zeta \in F'$;

(v) $(x^e\zeta-1)(f_2-1)...(f_k-1)$

$\equiv (x^e-1)(f_2-1)...(f_k-1) - ([x^e,f_2,...,f_k] - 1) \mod (\mathfrak{f}\mathfrak{a} + \mathfrak{R}(k)^*)$, *for all* $f_i,x \in F$, $\zeta \in F'$ *and* $x^e\zeta \in R$.

(vi) $(r-1)(f_2-1)(f_3-1)...(f_k-1) \equiv (f_2-1)(r-1)(f_3-1)...(f_k-1) +$

$$\sum_{i=3}^{k} (f_i-1)[(r-1)(f_2-1) - (f_2-1)(r-1)](f_3-1)...(f_{i-1}-1)(f_{i+1}-1)...(f_k-1)$$

mod $(\mathfrak{f}^2\mathfrak{a} + \mathfrak{R}(k)^* + \mathfrak{r}(k+1))$, *for all* $f_i \in F$, $r \in R$.

As an immediate consequence of Lemma 7.5 (i)-(v), we deduce that modulo

$$\mathfrak{f}\mathfrak{a} + \mathfrak{R}(k)^* + \mathfrak{f}^n,$$

the ideal $\mathfrak{r}(k)$, $k \ge 3$, is spanned by all elements of the form

$$(x_1-1)^{q(1)} \ldots (x_i^{e(i)}\zeta_i-1)(x_i-1)^{q(i)-1} \ldots (x_m-1)^{q(m)} \tag{1}$$

where each $q(j) \ge 0$, $j \ne i$, $q(i) \ge 1$ and $k \le q(1) + ... + q(m) \le n-1$.

We note by Lemma 7.5 (iv) that if either $i \ge 2$ and $q(j) \ne 0$ for some $j < i$, or $i = 1$ and $q(i) \ge 2$ then, modulo $\mathfrak{f}\mathfrak{a}$, (1) becomes

$$(x_1-1)^{q(1)} \ldots (x_i^{e(i)}-1)(x_i-1)^{q(i)-1} \ldots (x_m-1)^{q(m)}. \tag{2}$$

It follows by (1) and (2) that, for $k \ge 3$, modulo $\mathfrak{f}\mathfrak{a} + \mathfrak{R}(k)^* + \mathfrak{f}^n$, $\mathfrak{r}(k)$ consists of $\mathbf{Z}$-linear sums of all elements u_i, $i = 1,...,m$, defined by

$$u_i = (x_i^{e(i)}\zeta_i-1)(x_{i+1}-1)^{q(i+1)} \ldots (x_m-1)^{q(m)} + y_i, \tag{3}$$

where $k \le 1 + q(i+1) + ... + q(m) \le n-1$ and y_i belongs to the ideal $\mathfrak{y}(i)$ defined by

$$\mathfrak{y}(i) = (x_i-1)\mathfrak{x}(i) + \mathfrak{x}(i+1) + \ldots + \mathfrak{x}(m),$$

where

$$\mathfrak{x}(i) = (x_i^{e(i)} - 1)\mathbf{Z}F. \tag{4}$$

For each u_i given by (3), we define

$$v_i = (x_i^{e(i)}-1)\,(x_{i+1}-1)^{q(i+1)} \ldots (x_m-1)^{q(m)}, \tag{5}$$

and the corresponding group commutator

$$g(v_i) = [x_i^{e(i)},\ q(i+1)x_{i+1},\ \ldots,\ q(m)x_m], \tag{6}$$

with x_j appearing $q(j)$ times.

By Lemma 7.5(v), it follows that, modulo $\mathfrak{f}\mathfrak{a} + \mathfrak{R}(k)^* + \mathfrak{f}^n$, each u_i of the form (3) can be written as

$$\begin{aligned} u_i &= (x_i^{e(i)}\zeta_i-1)(x_{i+1}-1)^{q(i+1)} \ldots (x_m-1)^{q(m)} + y_i \\ &\equiv (x_i^{e(i)}-1)(x_{i+1}-1)^{q(i+1)} \ldots (x_m-1)^{q(m)} \\ &\qquad + ([x_i^{e(i)},\ q(i+1)x_{i+1},\ \ldots,\ q(m)x_m]^{-1} - 1) + y_i \\ &\equiv v_i + (g(v_i)^{-1}-1) + y_i. \end{aligned}$$

Further, since

$$u_i = (x_i^{e(i)}\zeta_i-1)(x_j-1)(x_j-1)^{q(j)-1} \ldots (x_m-1)^{q(m)} + y_i,$$

for some $i < j$ and $q(j) \geq 1$, we have the following characterization of elements of $\mathfrak{r}(k)$, $k \geq 3$.

Lemma 7.6. *Let* u *be an arbitrary element of* $\mathfrak{r}(k)$, $k \geq 3$. *Then modulo*

$$\mathfrak{f}\mathfrak{a} + \mathfrak{R}(k)^* + \mathfrak{f}^n,$$

u *is of the form*

$$u = \sum_{1\leq i<j\leq m} v(i,j) + (g(v(i,j))^{-1}-1) + y(i,j),$$

where, by (5) (6) and (4),

$$v(i,j) = (x_i^{e(i)}-1)(x_j-1)p(i,j); \tag{7}$$

$$g(v(i,j)) = [x_i^{e(i)},x_j]^{p(i,j)}; \tag{8}$$

with $p(i,j) = p(x_j,...,x_m) \in \mathfrak{f}^{k-2}$; *and* $y(i,j) \in \mathfrak{y}(i,j) + \mathfrak{a} + \mathfrak{f}^n$, *where*

$$\mathfrak{y}(i,j) = (x_i-1)\mathfrak{x}(i) + \mathfrak{x}(j) + \mathfrak{x}(j+1) + ... + \mathfrak{x}(m). \tag{9}$$

Conversely, modulo $\mathfrak{f}\mathfrak{a} + \mathfrak{R}(k)^* + \mathfrak{f}^n$, each $v(i,j) +(g(v(i,j))^{-1}-1) + y(i,j)$ lies in $\mathfrak{r}(k)$, $k \geq 2$, whenever $v(i,j)$, $g(v(i,j))$ are defined by (7), (8) and

$$y(i,j) \in \mathfrak{y}(i,j) + \mathfrak{a} + \mathfrak{f}^n$$

defined by (9).

Using Lemma 7.6 we can now deduce the following characterization of the subgroup

$$F \cap (1 + \mathfrak{r}(k) + \mathfrak{f}\mathfrak{a} + \mathfrak{f}^n)$$

for $k \geq 3$.

Theorem 7.7. *For* $k \geq 3$, $F \cap (1 + \mathfrak{r}(k) + \mathfrak{f}\mathfrak{a} + \mathfrak{f}^n) = G(k)R(k)F''\gamma_n(F)$, *where* $G(k)$ *is the subgroup generated by all elements of the form*

$$g(v) = \prod_{1\leq i<j\leq m} g(v(i,j)) = \prod_{1\leq i<j\leq m} [x_i^{e(i)}, x_j]^{p(i,j)}, \tag{10}$$

where $p(i,j) = p(x_j,...,x_m) \in \mathfrak{f}^{k-2}$ *are such that*

$$v = \sum_{1\leq i<j\leq m} v(i,j)$$

$$= \sum_{1\leq i<j\leq m} (x_i^{e(i)}-1)(x_j-1)p(i,j) + y(i,j) \equiv 0 \bmod \mathfrak{a} + \mathfrak{f}^n, \tag{11}$$

for some $y(i,j) \in \mathfrak{y}(i,j)$ *as defined by (9).*

We conclude by deducing, as a corollary, the following special case of Theorem B.

Theorem 7.8. *For* $m=2$ *and* $k \geq 3$, $F \cap (1 + \mathfrak{r}(k) + \mathfrak{f}\mathfrak{a} + \mathfrak{f}^n) = R(k)F''\gamma_n(F)$.

Proof. Let $g(v)$ be as in Theorem 7.7. Then, since $m=2$ and $k \geq 3$,

$$g(v) = [x_1^{e(1)},x_2]^{p(1,2)},$$

where $p(1,2) = p(x_2) = (x_2-1)p^*(1,2)$ is such that

$$(x_1^{e(1)}-1)(x_2-1)^2p^*(1,2) \equiv 0 \bmod (x_1-1)\mathfrak{x}(1) + \mathfrak{x}(2) + \mathfrak{a} + \mathfrak{f}^n. \tag{12}$$

Dividing (12) by (x_1-1) gives

$$t_1(x_1)(x_2-1)^2p^*(1,2) \equiv 0 \bmod \mathfrak{x}(1) + \mathfrak{x}(2) + \mathfrak{a} + \mathfrak{f}^{n-1}, \tag{13}$$

where $t_1(x_1) = 1 + x_1 + ... + x_1^{e(1)-1}$.

Dividing (13) by (x_2-1) gives

$$t_1(x_1)(x_2-1)p^*(1,2) \equiv 0 \bmod \mathfrak{x}(1) + t_2(x_2)ZF + \mathfrak{a} + \mathfrak{f}^{n-2}. \tag{14}$$

Since the left hand side of (14) is further divisible by (x_2-1), (14) becomes

$$t_1(x_1)(x_2-1)p^*(1,2) \equiv 0 \bmod \mathfrak{x}(1) + \mathfrak{x}(2) + \mathfrak{a} + \mathfrak{f}^{n-2},$$

which is the same as

$$t_1(x_1)p(1,2) \equiv 0 \bmod \mathfrak{x}(1) + \mathfrak{x}(2) + \mathfrak{a} + \mathfrak{f}^{n-2}. \tag{15}$$

Using (15) we have modulo $R(k)F''\gamma_n(F)$,

$$\begin{aligned} g(v) &= [x_1^{e(1)},x_2]^{p(1,2)} \\ &\equiv [x_1,x_2]^{t_1(x_1)p(1,2)} \\ &\equiv 1, \end{aligned}$$

since $[x_1,x_2]^{\mathfrak{a} + \mathfrak{f}^{n-2}} \leq F''\gamma_n(F)$ and since, for $s = 1,2$, $y_i \in F$,

$$[x_1,x_2,x_s^{e(s)},y_4,...,y_k] \in R(k)F''.$$

This completes the proof of the theorem.

Remark. Theorem 7.8 yields that if G is a 2-generated metabelian odd p-group then $D_n(G) = \gamma_n(G)$ for all n. The proof of the general case requires deeper analysis of the structure of elements of $\mathfrak{r}(k) + \mathfrak{f}\mathfrak{f}\mathfrak{a} + \mathfrak{f}^n$. This is given in Gupta (1989).

8. Other aspects and problems

Apart from what has been reported in the previous sections, there is very little known towards the solution of the general dimension subgroup problem. Recall Hall-Jennings theorem that if the lower central quotients of a group G are all torsion free then $D_n(G) = \gamma_n(G)$ for all n. Free solvable groups fall in this category and so the dimension subgroup conjecture remains valid for these groups. If, however, G is a free centre-by-metabelian group then $\gamma_i(G)/\gamma_{i+1}(G)$ are known to have elementary abelian 2-groups as subgroups (Ridley 1970, Hurley 1972), and as such are not covered by the Hall-Jennings result. Thus in all such cases a different approach is needed and the problem may turn out to be unusually complicated. For free centre-by-metabelian groups the problem has been resolved affirmatively by C K Gupta and Levin (1985). We give an outline of the technique used. The language used is that of free group rings.

Let $R = [F'',F]$ be the subgroup of F generated by all commutators

$$[[f_1,f_2], [f_3,f_4],f_5],\ f_i \in F.$$

Then F/R is a free centre-by-metabelian group. It is easy to verify that

$$[[f_1,f_2],[f_3,f_4],f_5] - 1 \in \mathfrak{faf},$$

so as a first approximation to the identification of $F \cap (1 + \mathfrak{r} + \mathfrak{f}^n)$, Gupta and Levin were able to identify the normal subgroup $F \cap (1 + \mathfrak{faf} + \mathfrak{f}^n)$ as follows:

Theorem 8.1 (C K Gupta and Levin (1985). *Let* $D(c) = F \cap (1 + \mathfrak{faf} + \mathfrak{f}^{c+1})$. *Then*

(i) $D(c) = \gamma_{c+1}(F)$ *if* $1 \le c \le 4$;

(ii) $F(c) = T_c(F)K_6(F)[F'',F]\gamma_{c+1}(F)$ *if* c *is odd*, $c \ge 5$;

(iii) $D(c) = T_{c-1}(F)U_c(F)K_6(F)[F'',F]\gamma_{c+1}(F)$ *if* c *is even*, $c \ge 6$,

where $U_c(F)$ *is the fully invariant closure of* $[f_1,f_2; f_1, f_2, ..., f_{c-2}]$ *and* $K_6(F)$, $T_c(F)$ *are fully invariant closures of products of certain specifically defined complex commutators.*

To prove

$$F \cap (1 + \mathfrak{r} + \mathfrak{f}^{c+1}) = [F'',F]\gamma_{c+1}(F)$$

it is necessary to eliminate each of the unwanted subgroups $U_c(F)$, $K_6(F)$, $T_c(F)$ which appear in the above identification and this is achieved through a series of steps. The details are too technical to report.

Recall that the counter-examples to the dimension subgroup conjecture illustrate that in general

$$F \cap (1 + \mathfrak{r} + \mathfrak{f}^n) \ne R\gamma_n(F)$$

for $n \ge 4$ even when F/R is assumed to be metabelian. In these counter-examples the prime 2 played an important role. Also recall that $F \cap (1 + \mathfrak{r} + \mathfrak{f}^n) = R\gamma_n(F)$ if

F/R is a metabelian p-group, p odd. In contrast, we next offer a construction of a metabelian p-group F/R, $p \geq 5$, such that

$$F \cap (1 + \mathfrak{r} + \mathfrak{f}\mathfrak{a} + \mathfrak{f}^5) \neq R\gamma_5(F).$$

[Note that $F \cap (1 + \mathfrak{r} + \mathfrak{f}^5) = R\gamma_5(F)$ (since p is odd and F/R is metabelian) and $F \cap (1 + \mathfrak{r}) = F \cap (1 + \mathfrak{r} + \mathfrak{f}\mathfrak{a}) = R$.]

Example. Let $F = \langle a,b \rangle$ be the free group and let $p \geq 5$ be a fixed prime. For $R = \langle a^p[a,b,b]^{-1}, b^p[b,a,a]^{-1} \rangle^F F''$,

$$F \cap (1 + \mathfrak{r} + \mathfrak{f}\mathfrak{a} + \mathfrak{f}^5) \neq R\gamma_5(F).$$

Details. It suffices to prove that

$$[a,b]^p \notin R\gamma_5(F) \text{ whereas } [a,b]^p - 1 \in \mathfrak{r} + \mathfrak{f}\mathfrak{a} + \mathfrak{f}^5.$$

Since modulo $R\gamma_5(F)$ we have the congruences:

$$[a^p,a] \equiv [a,b,b,a] \equiv 1; \qquad [a^p,b] \equiv [a,b,b,b];$$

$$[b^p,a] \equiv [b,a,a,a]; \qquad [b^p,b] \equiv [b,a,a,b] \equiv 1;$$

$$[a^p,b] \equiv [a,b]^p; \qquad [b^p,a] \equiv [b,a]^p \quad (\text{since } p \geq 5);$$

$$[a,b,b,b] \equiv [b,a,a,a]^{-1}.$$

Thus $[a,b]^p \notin R\gamma_5(F)$. On the other hand, using the fact that

$$(a^p - 1) \equiv p(a-1) \text{ modulo } \mathfrak{f}^4 \ (\text{since } p \geq 5),$$

we have the following congruences modulo $\mathfrak{r} + \mathfrak{f}\mathfrak{a} + \mathfrak{f}^5$,

$$\begin{aligned}
[a,b]^p - 1 &\equiv p([a,b] - 1) \\
&\equiv p(a-1)(b-1) - p(b-1)(a-1) \\
&\equiv (a-1)(b^p-1) - (b-1)(a^p-1) \\
&\equiv (a-1)(b^p[b,a,a]^{-1}-1) - (b-1)(a^p[a,b,b]^{-1}-1) \\
&\equiv 0.
\end{aligned}$$

This completes the details of the construction.

Since for F/R a metabelian p-group, p odd,

$$F \cap (1 + \mathfrak{r} + \mathfrak{a}^2 + \mathfrak{f}^5) = RF''\gamma_5(F) \text{ (Gupta 1989)}$$

and since

$$F \cap (1 + \mathfrak{f}\mathfrak{a} + \mathfrak{f}^5) = F''\gamma_5(F) = F \cap (1 + \mathfrak{a}^2 + \mathfrak{f}^5) \text{ (Gupta 1982),}$$

with $\mathfrak{x} = \mathfrak{f}\mathfrak{a} + \mathfrak{f}^5$, $\mathfrak{y} = \mathfrak{a}^2 + \mathfrak{f}^5$ the above example serves to illustrate the following interesting phenomenon:

$$F \cap (1 + \mathfrak{x} + \mathfrak{f}^5) \neq F \cap (1 + \mathfrak{y} + \mathfrak{f}^5)$$

whereas

$$F \cap (1 + \mathfrak{x}) = F \cap (1 + \mathfrak{y}).$$

Extending the identification $F \cap (1 + \mathfrak{f}\mathfrak{a} + \mathfrak{f}^n) = F''\gamma_n(F)$, Gupta-Gupta-Levin (1987) have proved that if $\mathfrak{a}_c = ZF(\gamma_c(F)\text{-}1)$, $c \geq 2$, then

$$F \cap (1 + \mathfrak{f}\mathfrak{a}_c + \mathfrak{f}^n) = [\gamma_c(F), \gamma_c(F)]\, \gamma_n(F).$$

We close with the following open problems:

Problem 5. Identify the subgroup $F \cap (1 + \mathfrak{f}\mathfrak{n} + \mathfrak{f}^n)$ when F/N is a finitely generated nilpotent group and $\mathfrak{n} = ZF(N\text{-}1)$.

Problem 6. Identify the following higher dimension subgroups:

(i) $F \cap (1 + \mathfrak{r}(k) + \mathfrak{f}^n)$, (ii) $F \cap (1 + \mathfrak{r}^k + \mathfrak{f}^n)$, $n > k \geq 2$,

where $\mathfrak{r}(k) = \sum_{i+j=k-1} \mathfrak{f}^i \mathfrak{r} \mathfrak{f}^j$.

Problem 7. Is the quotient group $F \cap (1 + \mathfrak{r} + \mathfrak{f}^n)/R\gamma_n(F)$ always abelian? [This is known to be the case when F/R is metabelian (Gupta and Passi) (see Gupta (1987), IV.3.7).]

Problem 8 (cf. Plotkin 1973). For $n \geq 4$, does there exist $m = m(n)$ such that

$$F \cap (1 + \mathfrak{r} + \mathfrak{f}^m) \leq R\gamma_n(F)$$

for all normal subgroups R of F?

References

F Bachman & L Grünenfelder, Homological methods and the third dimension subgroup, *Comment. Math. Helv.* **47** (1972), 526-531.

S Bachmuth, Automorphism groups of free metabelian groups, *Trans. Amer. Math. Soc.* **118** (1965), 93-104.

Bruce Chandler & Wilhelm Magnus, *The history of combinatorial group theory: A case study in the history of ideas* (Springer-Verlag, 1982).

K T Chen, R H Fox & R C Lyndon, Free differential calculus IV: The quotient groups of the lower central series, *Ann. of Math.* **68** (1958), 81-95.

G Cliff & B Hartley, Sjogren's theorem on dimension subgroups, *J. Pure Appl. Algebra* **47** (1987), 231-242.

P M Cohn, Generalization of a theorem of Magnus, *Proc. London Math. Soc.* (3) **2** (1952), 297-310.

Ralph H Fox, Free differential calculus I - Derivations in free group rings, *Ann. of Math.* **57** (1953), 547-560.

O Grün, Über eine Fakturgruppe freier Gruppen I, *Deutsche Mathematik* **1** (1936), 772-782.

O Grün, Zusammenhang zwischen Potenzbildung und Kommutatorbildung, *J. Reine Angew. Math.* **182** (1940), 158-177.

K W Gruenberg, Residual properties of infinite soluble groups, *Proc. London Math. Soc.* (3) **7** (1957), 29-62.

C K Gupta, N D Gupta & F Levin, *On dimension subgroups relative to certain product ideals* (Springer Lecture Notes in Math. **1281**, Berlin, New York 1987), 31-35.

Chander Kanta Gupta & Frank Levin, Dimension subgroups of free centre-by-metabelian groups, *Illinois J. Math.* **30** (1986), 258-273.

Narain Gupta, On the dimension subgroups of metabelian groups, *J. Pure Appl. Algebra* **24** (1982), 1-6.

Narain Gupta, Sjogren's theorem for dimension subgroups - the metabelian case, *Ann. of Math. Stud.* **111** (1987), 197-211.

Narain Gupta, Free group rings, *Contemp. Math.* **66** (1987).

Narain Gupta, Dimension subgroups of metabelian p-groups, *J. Pure Appl. Algebra* **51** (1988), 241-249.

Narain Gupta, A solution of the dimension subgroup problem, *J. Algebra* (1989), to appear.

Narain Gupta, The dimension subgroup conjecture, *Bull. London Math. Soc.*, to appear.

Narain Gupta & Ken-Ichi Tahara, Dimension and lower central subgroup of metabelian p-groups, *Nagoya Math. J.* **100** (1985), 127-133.

N D Gupta, A W Hales & I B S Passi, Dimension subgroups of metabelian groups, *J. Reine Angew. Math.* **346** (1984), 194-198.

Alfred W Hales, Stable augmentation quotients of abelian groups, *Pacific J. Math.* **118** (1985), 401-410.

B Hartley, *Dimension and lower central subgroups - Sjogren's theorem revisited* (Lecture Notes **9**, National University of Singapore, 1982).

A H M Hoare, Group rings and lower central series, *J. London Math. Soc.* (2) **1** (1969), 37-40.

T C Hurley, Representations of some relatively free groups in power series rings, *Proc. London Math. Soc.* (3) **24** (1972), 257-294.

T C Hurley & S K Sehgal, The Lie dimension subgroup conjecture, *J. Algebra* (1989), to appear.

S A Jennings, The structure of a group ring of a p-group over a modular field, *Trans. Amer. Math. Soc.* **50** (1941), 175-185.

Gerald Losey, On dimension subgroups, *Trans. Amer. Math. Soc.* **97** (1960), 474-486.

Gerald Losey, N-series and filtrations of the augmentation ideal, *Canad. J. Math.* **26** (1974), 962-977.

Roger Lyndon, Problems in Combinatorial Group Theory, *Ann. of Math. Stud.* **111** (1987), 3-33.

Wilhelm Magnus, Beziehungen zwischen Gruppen und Idealen in einem speziellen Ring, *Math. Ann.* **111** (1935), 259-280.

Wilhelm Magnus, Über Beziehunger zwischen höheren Kommutatoren, *J. Reine Angew. Math.* **177** (1937), 105-115.

Wilhelm Magnus, On a theorem of Marshall Hall, *Ann. of Math. (Ser. II)* **40** (1939), 764-768.

Wilhelm Magnus, Abraham Karrass & Donald Solitar, *Combinatorial Group Theory* (Interscience, 1966, New York).

S Moran, Dimension subgroups mod n, *Proc. Cambridge Philos. Soc.* **68** (1970), 579-582.

I B S Passi, Dimension subgroups, *J. Algebra* **9** (1968), 152-182.

Inder Bir S Passi, Polynomial maps on groups, *J. Algebra* **9** (1968), 121-151.

Inder Bir S Passi, Polynomial maps on groups II, *Math. Z.* **135** (1974), 137-141.

Inder Bir S Passi, *Group Rings and their Augmentation Ideals* (Lecture Notes in Math. **715**, Springer Verlag, 1979).

I B S Passi, The free group ring, in: *Algebra and its applications* (Lecture Notes in Pure Appl. Math. **91**, 1989, Marcel Dekker).

Inder Bir S Passi & Sudarshan K Sehgal, Lie dimension subgroups, *Comm. Algebra* **3** (1975), 59-73.

I B S Passi, Sucheta & Ken-Ichi Tahara, Dimension subgroups and Schur multiplicator III, *Japan J. Math.* **13** (1987), 371-379.

B I Plotkin, Remarks on stable representation of nilpotent groups, *Trans. Moscow Math. Soc.* **29** (1973), 185-200.

J N Ridley, The free centre-by-metabelian group of rank two, *Proc. London Math. Soc.* (3) **20** (1970), 321-347.

E Rips, On the fourth integer dimension subgroups, *Israel J. Math.* **12** (1972), 342-346.

Frank Röhl, Review and some critical comments on a paper of Grün concerning the dimension subgroup conjecture, *Bol. Soc. Brasil Mat.* **16** (1985), 11-27.

Robert Sandling, The dimension subgroup problem, *J. Algebra* **21** (1972), 216-231.

Robert Sandling, Dimension subgroups over arbitrary coefficient rings, *J. Algebra* **21** (1972), 250-265.

J A Sjogren, Dimension and lower central subgroups, *J. Pure Appl. Algebra* **14** (1979), 175-194.

John R Stallings, Quotients of powers of the augmentation ideal, *Ann. of Math. Stud.* **84** (1975), 101-118.

Ken-Ichi Tahara, On the structure of $Q_3(G)$ and the fourth dimension subgroup, *Japan J. Math. (NS)* **3** (1977), 381-396.

Ken-Ichi Tahara, The augmentation quotients of group rings and the fifth dimension subgroups, *J. Algebra* **71** (1981), 141-173.

E Witt, Treue Darstellung Liesher Ringe, *J. Reine Angew. Math.* **177** (1937), 152-160.

Printed in the United States
By Bookmasters